FOCUS ON

Grades 5-8

MIDDLE SCHOOL

ASTRONOMY

Laboratory Notebook

3rd Edition

Rebecca W. Keller, PhD

REAL SCIENCE 4 Kids

Real Science-4-Kids

Illustrations: Janet Moneymaker

Focus On Middle School Astronomy Laboratory Notebook—3rd Edition
ISBN 978-1-941181-46-1

Published by Gravitas Publications Inc.
www.gravitaspublications.com
www.realscience4kids.com

GRAVITAS
PUBLICATIONS

Keeping a Laboratory Notebook

A laboratory notebook is essential for the experimental scientist. In this type of notebook, the results of all the experiments are kept together along with comments and any additional information that is gathered. For this curriculum, you should use this workbook as your laboratory notebook and record your experimental observations and conclusions directly on its pages, just as a real scientist would.

The experimental section for each chapter is pre-written. The exact format of a notebook may vary among scientists, but all experiments written in a laboratory notebook have certain essential parts. For each experiment, a descriptive but short *Title* is written at the top of the page along with the *Date* the experiment is performed. Below the title, an *Objective* and a *Hypothesis* are written. The objective is a short statement that tells something about why you are doing the experiment, and the hypothesis is the predicted outcome. Next, a *Materials List* is written. The materials should be gathered before the experiment is started.

Following the *Materials List,* the *Experiment* is written. The sequence of steps for the experiment is written beforehand, and any changes should be noted during the experiment. All of the details of the experiment are written in this section. All information that might be of some importance is included. For example, if you are to measure 236 ml (1 cup) of water for an experiment, but you actually measured 300 ml (1 1/4 cup), this should be recorded. It is hard sometimes to predict the way in which even small variations in an experiment will affect the outcome, and it is easier to track down a problem if all of the information is recorded.

The next section is the *Results* section. Here you will record your experimental observations. It is extremely important that you be honest about what is observed. For example, if the experimental instructions say that a solution will turn yellow, but your solution turned blue, you must record blue. You may have done the experiment incorrectly, or you might have discovered a new and interesting result, but either way, it is very important that your observations be honestly recorded.

Finally, the *Conclusions* should be written. Here you will explain what the observations may mean. You should try to write only valid conclusions. It is important to learn to think about what the data actually show and what cannot be concluded from the experiment.

Contents

Experiment 1

Constellations

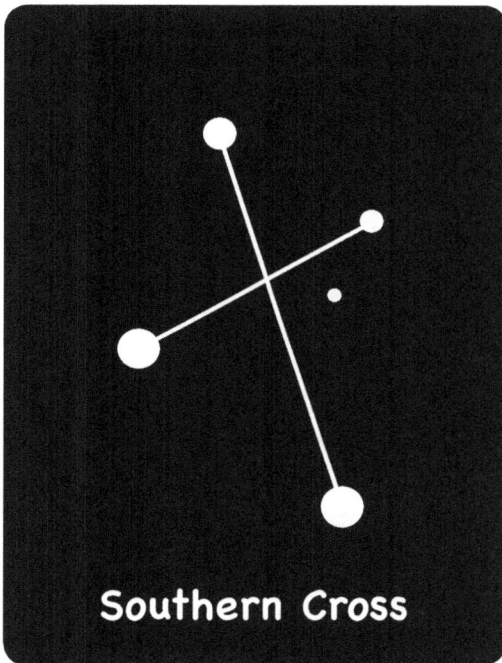

Southern Cross

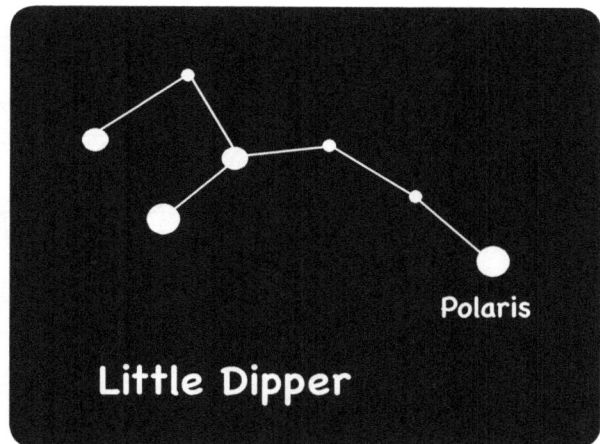

Little Dipper

Polaris

Introduction

Find some constellations and see if you can use the stars to tell which direction you are going.

I. Think About It

❶ Do you think it was important for ancient people to be able to recognize different stars? Why or why not?

_____

_____

_____

_____

_____

❷ Why do think a star is in a different place in the sky in the morning than at night?

_____

_____

_____

_____

_____

❸ Do you think stars move from one constellation into another? Why or why not?

_____

_____

_____

_____

_____

❹ Do you think people are discovering new constellations all the time? Why or why not?

_____

_____

_____

_____

❺ If you are in the Northern Hemisphere and you use the North Star to find north, how would you find south?

_____

_____

_____

_____

❻ Do you think sailors were able to find their way at sea before compasses were invented? Why or why not?

_____

_____

_____

_____

II. Experiment 1: Constellations

Date_____

Objective _____

_____

Hypothesis _____

_____

_____

Materials

pencil
flashlight
compass

A clear night sky away from bright lights is needed.

EXPERIMENT

Record your physical location, city, state, or country, whether you are in the Northern or Southern Hemisphere, and the month.

Northern Hemisphere

Location	
Hemisphere	
Month	

❶ In the evening on a clear night away from city lights go outside and, without using a compass, locate "north." To do this you will need to find the Big Dipper. The Big Dipper is a set of stars that form the shape of a "dipping spoon." (The Big Dipper is not an official constellation but is called an *asterism* — a small group of stars.) The two stars on the end of the dipping spoon point to the star Polaris.

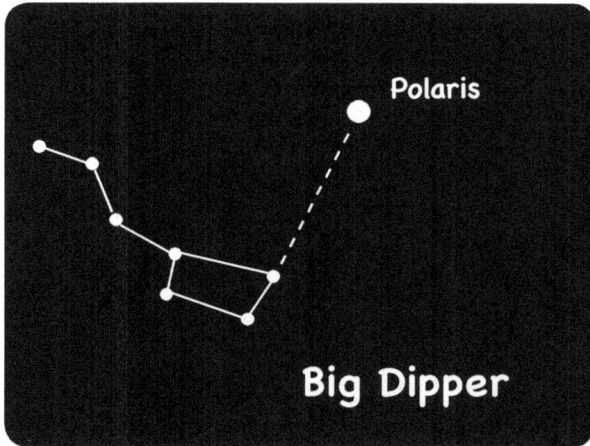

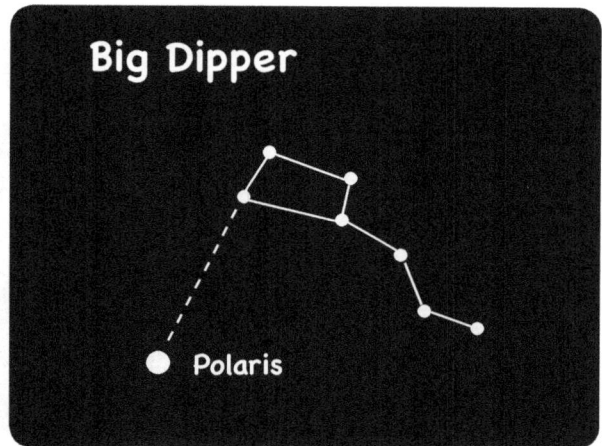

Polaris is the "North Star," and when you turn towards Polaris, you are pointing "north." It doesn't matter how the Big Dipper is oriented in the sky, the two end stars always point to the North Star. The North Star is the only star in the sky that doesn't move (much). All of the constellations appear to move around the North Star. Once you find the North Star you can find nearby constellations.

❷ Now that you have found the North Star, try to find the constellation called the "Little Dipper."

Polaris forms the end of the handle of the Little Dipper.

❸ In the following box, draw the Little Dipper as it looks to you.

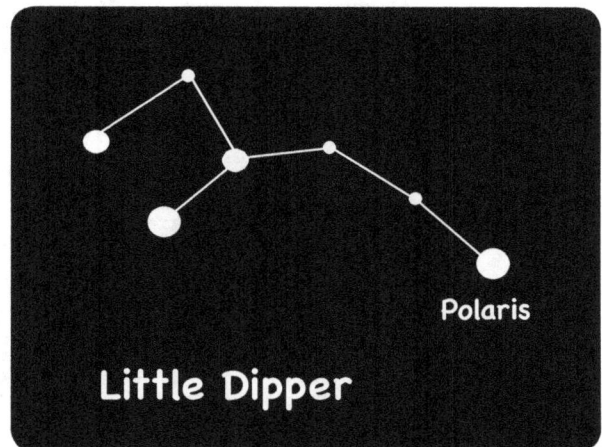

Little Dipper

❹ Try to locate the "Dragon." The Dragon constellation is between the Big Dipper and Little Dipper.

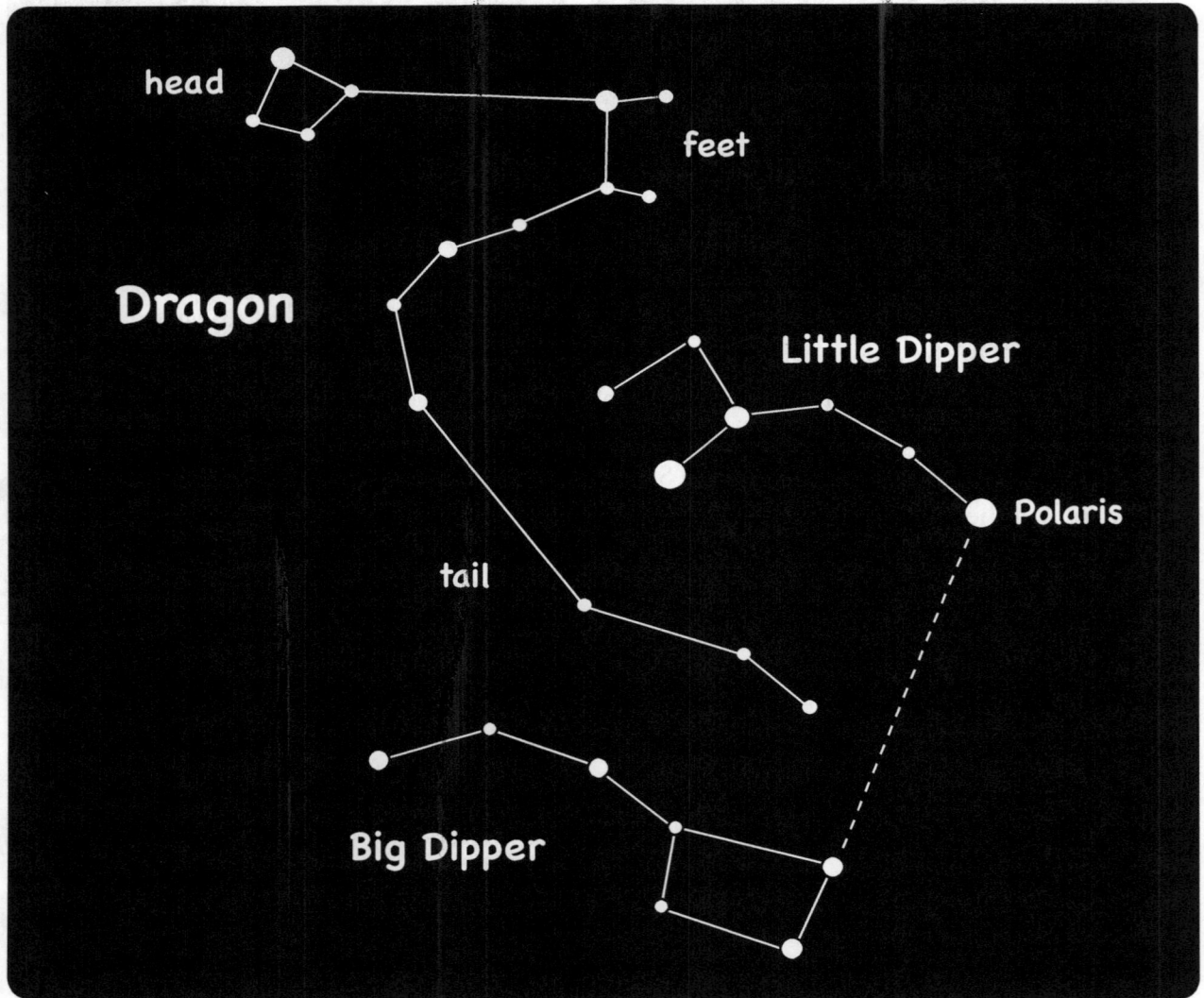

❺ On the following page, draw the Dragon constellation as you see it.

Draw the Dragon constellation as you observe it.

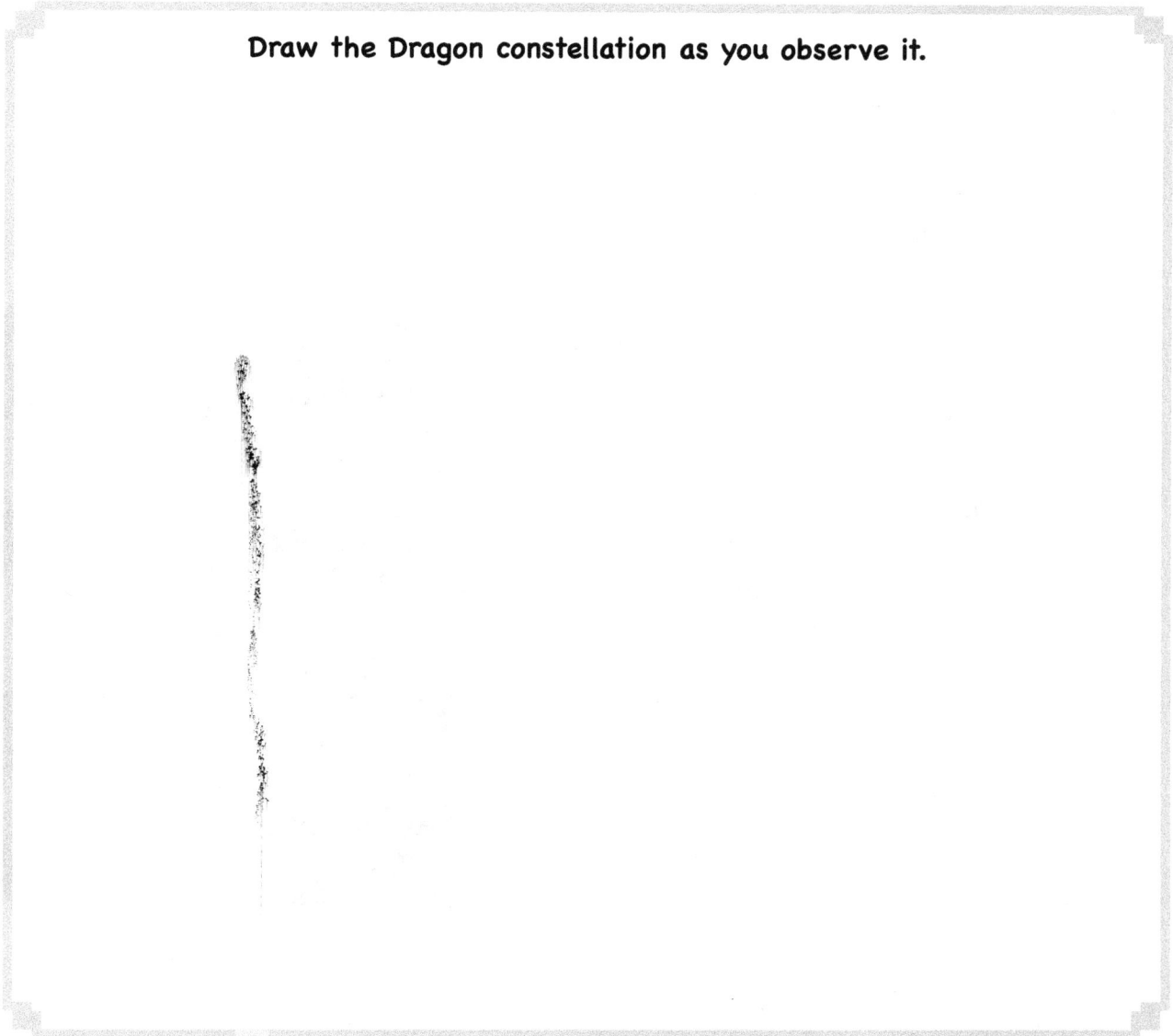

❻ Count the stars in the Dragon constellation in the image on the previous page. Compare this number with the number of stars you've recorded for the Dragon. Are they the same? Why or why not?

_____

_____

_____

❼ Find the North Star again and use your compass to see if the North Star really is above the North Pole.

EXPERIMENT

Southern Hemisphere

The South Pole doesn't have a star directly over it like the North Pole does, but you can still find south using the stars.

❶ In the evening on a clear night away from city lights go outside. Look toward the south to find the Southern Cross constellation, also called Crux. It is a small, bright constellation of four stars that are close together. You may see two crosses near each other. The Southern Cross is smaller and has brighter stars. it also has a dimmer fifth star tucked in between two of its arms. The larger, dimmer cross is called the False Cross and is an asterism rather than a constellation.

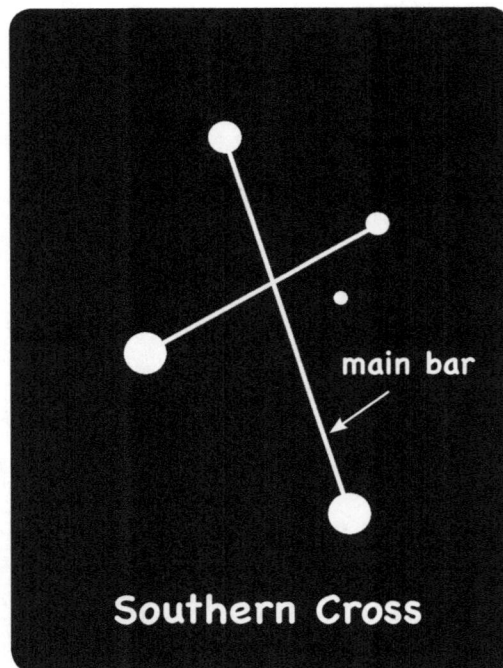

❷ While looking at the Southern Cross, follow with your eyes the main bar of the cross (the line between the two stars that are farthest apart).

❸ Now imagine you are extending the main bar downward and adding 4 1/2 times to its length. By doing this you will arrive at a point in the sky called the South Celestial Pole which is directly above the South Pole.

❹ From the South Celestial Pole, lower your eyes straight down to the horizon. The point you are looking at will be south.

❺ Using your compass, see how close you came to finding south.

❻ Record your results on the next page.

Results – Finding south by using the Southern Cross

_____

_____

_____

_____

_____

_____

_____

❼ See if you can find both the Southern Cross and the False Cross and then draw them as you see them.

The Southern Cross and the False Cross

III. Conclusions

Summarize how easy or difficult it was to find the constellations you were looking for. Do you think you could use these stars for navigation? What role, if any, does your physical location and the month you made these observations have on your results? What did you observe about the night sky that you hadn't noticed before?

_____

_____

_____

_____

_____

_____

_____

_____

_____

_____

_____

_____

_____

IV. Why?

In ancient times people were very observant of the world around them. There were no city lights, so they could see the stars very clearly. They noticed that there was one star that didn't seem to change position over the course of the night and that all the other stars seemed to rotate around it. This star is now called the North Star.

Before the invention of the compass, people in the Northern Hemisphere were able to determine in which direction they were traveling at night by looking at the position of the North Star. If it was directly in front of them, they were going north; directly behind them, they were going south; to the right of them, they were going west; and to the left of them, they were going east.

Although the Southern Cross is not directly over the South Pole, ancient people discovered how it could be used in a similar manner to determine which way was south. Then the other directions could be found.

The International Astronomical Union (IAU) is an organization that holds meetings where astronomers from all over the world can get together to share ideas and research. The IAU is also the organization that gives official names to celestial bodies that are discovered. The IAU decided that it would be helpful to have an official set of constellations and in 1930 came up with the current list of 88 constellations. Half of these constellations come from the ancient Greeks who described them long ago.

V. Just For Fun

Use online or library resources to find more constellations that you can see from the area where you live. Pick your favorite three constellations. Go outside on a clear night, find the constellations, and draw them as you see them.

Favorite Constellations

Experiment 2

Measuring Distances

Mars Rover   Image Credit: NASA/JPL/Cornell University

Introduction

Use a simple triangulation method to measure the distance of a faraway object.

I. Think About It

❶ Do you think ancient astronomers knew the distance to different stars?
Why or why not?

_____

_____

_____

_____

❷ Do you think simple tools could be used to measure the distance to faraway objects?
Why or why not?

_____

_____

_____

_____

❸ Do you think the development of advanced technology has helped today's astronomers
find out the distance to different stars? Why or why not?

_____

_____

_____

_____

❹ Do you think you could measure the distance to a faraway object without using mathematical formulas? Why or why not?

_____

_____

_____

_____

_____

❺ Do you think a knowledge of mathematics is important to doing astronomy? Why or why not?

_____

_____

_____

_____

_____

❻ Do you think the invention of space telescopes has helped astronomers find out the distance to different stars? Why or why not?

_____

_____

_____

_____

_____

II. Experiment 2: Measuring Distances   Date _____

Objective _____

_____

Hypothesis _____

_____

_____

Materials

two sticks (used for marking)
two rulers
tape
string, several meters long (several yards)
protractor
square grid or graph paper (included in this chapter)

EXPERIMENT

❶ Find a wide open space with a distant object. The space can be a field, a city street, or even your own backyard.

❷ Pick two observation points, and place the sticks at these points. Mark one observation point "A" and the other "B."

❸ Take two rulers and tape them together at one end, making a right angle.

❹ Place the corner of the double ruler on observation point "A" with one end pointing towards the object you want to measure and the other end pointing towards observation point "B."

❺ Attach the string to the stick at observation point "A," and stretch it out along the side of the double ruler pointing towards observation point "B." The string will be used as a guide so that you walk in a straight line.

☆✩☼☾✩☆✩☼☾✩☆✩☼☾✩☆✩☼☾✩☆✩☼☾✩☆✩☼☾✩☆✩☼☾✩☆✩☼☾✩☆✩☼☾☆

❻ Holding the string, walk heel-to-toe from observation point "A" to observation point "B," making sure the string is still pointing at a 90 degree angle in the direction of point "B." Count your steps. Each step will equal one "foot." (Note: You may need to adjust the location of point "B" to maintain the right angle at point "A.")

❼ When you get to point "B," attach the string to the stick. Check to make sure the string is still pointing in the same direction as the ruler. In the *Results* section, record the number of steps between point "A" and point "B."

❽ From observation point "B" find the object whose distance you want to measure. Place the protractor on the string so that you can measure the angle between point "B" and the distant object.

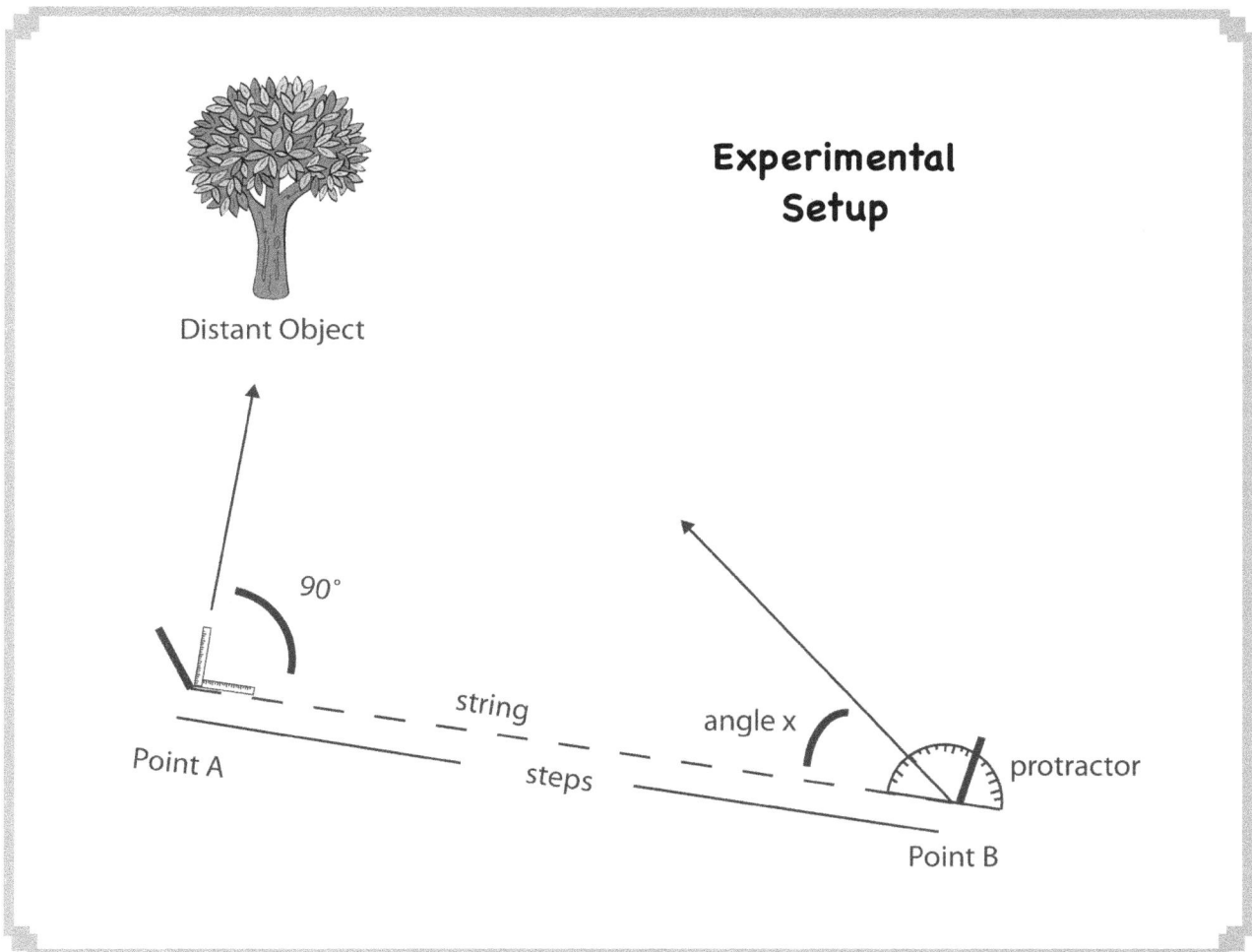

❾ Record the angle between point "B" and the distant object.

Distant Object

Experimental Setup

90°

string

angle x

Point A

steps

protractor

Point B

Results

Now that you have collected your data, use grid paper and a modeling technique to measure the distance to the object.

❶ On the grid paper on the following page, mark point "A."

❷ Using one square for each step, mark point "B" on the grid.

❸ Draw a line from point "A" to point "B." This is line "AB."

❹ Draw a line from point "A" towards the distant object. This line should be at a 90 degree angle to line AB. Label this line "y."

❺ Using your protractor, at point "B" on your graph mark the angle you measured at point "B." Draw a line from point "B" to the distant object using this angle. Extend this line until it intersects with line "y." (You may have to extend line "y" in order for the two lines to intersect.)

❻ Count the number of squares from point "A" along line "y" to the distant object.

❼ Assuming that each of your steps is one foot, how far away is the distant object? Record your answers below.

Number of steps — Point A to Point B _____

Angle at Point B _____

Number of squares — Point A to distant object _____

Distance of object in feet _____

III. Conclusion

What conclusions can you draw from your observations?

Summarize how easy or difficult it was to measure the distance of a faraway object. Write down any problems or sources of error you might have noticed.

_____

_____

_____

_____

_____

_____

_____

_____

_____

_____

_____

_____

_____

IV. Why?

In this experiment you were able to calculate the distance of a faraway object by using two points, some basic geometry, graph paper, and a method called *triangulation*. Triangulation uses the concept of *similar triangles* to estimate distances. In geometry, similar triangles are those that have exactly the same shape and only differ in size. Because the three angles in a triangle always add up to 180 degrees, if we know the sizes of two of the angles, we can add them together and subtract the total from 180 degrees to get the size of the third angle. So, if two angles of one triangle are the same as two angles in another triangle, the third angle in both triangles will be the same. This is true even if the triangles are different sizes.

In this experiment you used triangulation by measuring the distance between two points (A and B), measuring the angles at A and B relative to the distant object, and using this information to draw a small triangle on a piece of graph paper. The drawn triangle is similar to (has the same angles as) the actual triangle you marked off with the three points (A, B, and the distant object). Because you used your feet to measure the actual distance between point A and point B, you know how far it is from point A to point B. By assigning a value of one foot to each square on the graph paper, you can then use the grid squares to estimate the distance between point A and the distant object by counting the number of squares between them on the graph paper.

In astronomy the technique of triangulation is called parallax. Centuries before the invention of the telescope, parallax was used by early astronomers to estimate distances between Earth and faraway planets and stars. Early astronomers could see a difference in the viewing angle of a star by looking at the star one day and then looking at it again months later. By measuring each of the two viewing angles, astronomers could calculate a distance to the faraway object. Using parallax measurements works well for stars that are closer than 400 light-years away. A light-year is the unit of distance that light will travel in one year. Since light travels at a velocity of about 300,000 km per second (186,411 mi./sec.), in one light-year the distance light travels is almost 10 trillion km or about 63,240 AU (astronomical units) with 1 AU being about 150 million km (93 million mi.) — the distance from Earth to the Sun.

V. Just For Fun

Part A. Find another distant object that you can easily walk to (for example, a pole or other object in the backyard, park, field, or city street). Choose an object that you can walk to in a straight line without needing to go around anything. Repeat the experiment. In the space below, record your data.

Part B. After you have made your graphed triangle, check your calculation by measuring the distance to the distant object by walking heel-to-toe and counting each step as one "foot." Record your results.

A. Calculated Distance to Object

Number of steps — Point A to Point B   _____

Angle at Point B   _____

Number of squares — Point A to distant object   _____

Calculated distance to object in "feet"   _____

B. Measured Distance to Object

Measured distance to object in "feet"   _____

Are the results of your calculation and your actual measurement the same or different? Why do you think you got this result?

_____

_____

_____

_____

_____

Lunar and Solar Eclipses

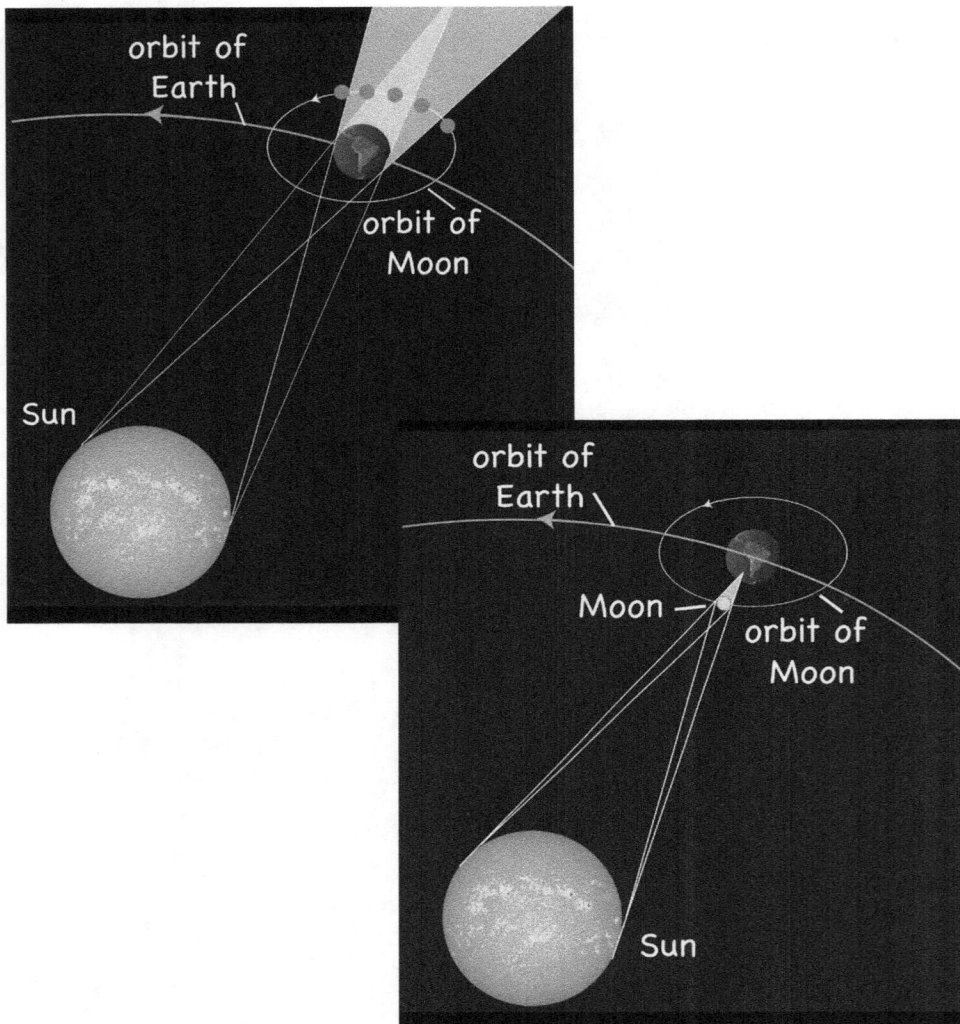

Introduction

Use a model to see how lunar and solar eclipses are created.

I. Think About It

❶ What do you think people learned about the Earth, Moon, and Sun by observing eclipses? Why?

_____

_____

_____

_____

_____

❷ Do you think there would be solar eclipses if Earth did not have a Moon? Why or why not?

_____

_____

_____

_____

❸ Do you think there would be eclipses if the Moon did not orbit Earth but always stayed in the same position above Earth? Why or why not?

_____

_____

_____

_____

❹ Do you think there is a lunar eclipse every time the Moon is on the opposite side of Earth from the Sun? Why or why not?

_____

_____

_____

_____

_____

❺ Do you think when there is a solar eclipse, everyone on the side of Earth facing the Sun will be able to see the solar eclipse? Why or why not?

_____

_____

_____

_____

_____

❻ If you were living at a space station on the Moon, do you think you would see any eclipses? Why or why not?

_____

_____

_____

_____

_____

II. Experiment 3: Lunar and Solar Eclipses   Date _____

Objective _____

_____

Hypothesis _____

_____

_____

Materials

basketball
ping-pong ball
flashlight
empty toilet paper tube
tape
scissors
a dark room

EXPERIMENT

In this experiment you will observe how lunar and solar eclipses occur.

❶ In a dark room, place the basketball on top of one end of a toilet paper tube that is sitting upright on the floor. The toilet paper tube will hold the basketball in place.

❷ Holding the flashlight, stand several feet away from the basketball. Turn on the flashlight and point it towards the basketball. Lay the flashlight on the floor in a position that keeps the basketball illuminated.

❸ Hold the ping-pong ball so that it is between the flashlight and the illuminated basketball. Adjust the position of the ping-pong ball until you can see its shadow on the basketball.

❹ Move the ping-pong ball up until there is no shadow on the basketball.

❺ Now lower the ping-pong ball until there is no shadow on the basketball.

❻ Move the ping-pong ball in an "orbit" around the basketball. Observe where the ping-pong ball needs to be in order for it to cast a shadow on the basketball and where the ping-pong ball needs to be for the basketball to cast a shadow on the ping-pong ball. Also note the position of the ping-pong ball when no shadows are cast.

Results

Repeat Step ❻ several times, moving the ping-pong ball in different orbits Note whether shadows are or are not cast.

Draw several of the "orbits" you test and note whether or not the ping-pong ball casts a shadow on the basketball or the basketball casts a shadow on the ping-pong ball (see the example below). You will need to spend some time "playing" with the ping-pong ball to find the positions where shadows occur.

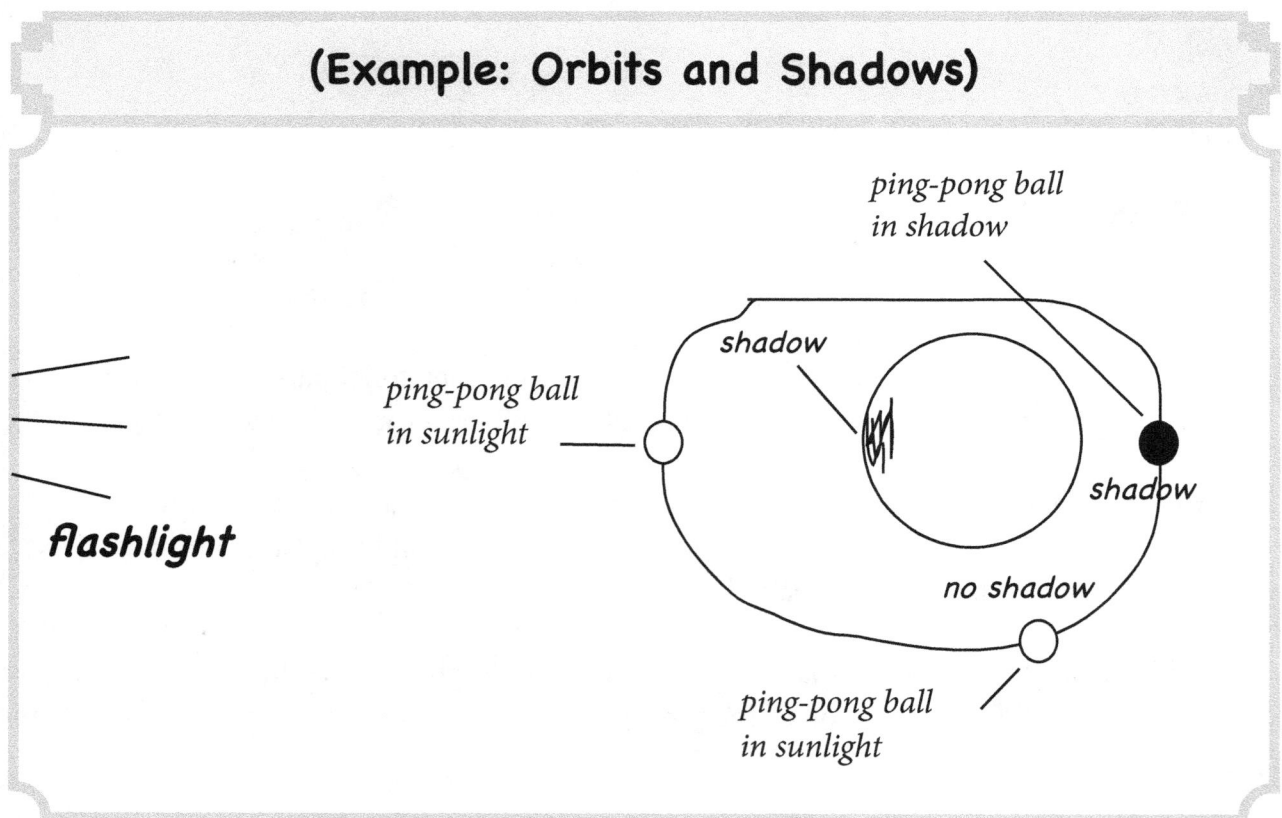

(Example: Orbits and Shadows)

ping-pong ball
in shadow

shadow

ping-pong ball
in sunlight

flashlight

shadow

no shadow

ping-pong ball
in sunlight

Orbits and Shadows

More Orbits and Shadows

III. Conclusion

Based on your observations, discuss how a lunar eclipse occurs.

_____

_____

_____

_____

_____

_____

_____

_____

Based on your observations, explain how a solar eclipse occurs.

_____

_____

_____

_____

_____

_____

_____

_____

IV. Why?

An eclipse is a fascinating event to witness. When the Earth passes in between the Sun and the Moon, a lunar eclipse occurs, and when the Moon passes between the Sun and Earth, we get a solar eclipse. Both types of eclipses can give us valuable information about the Earth, Moon, and Sun.

In order for an eclipse to occur, the Sun, Earth, and Moon have to be precisely lined up. If they are not, there won't be an eclipse, or the eclipse will be partial. In a partial lunar eclipse, the shadow of Earth moves over only a part of the Moon. In a partial solar eclipse, only part of the Sun will be covered as the Moon passes in front of it.

Because the Moon's orbit around the Earth is a little bit tipped when compared to Earth's orbit around the Sun, the Moon, Earth, and Sun don't always line up precisely enough to create an eclipse. This is why there isn't a total eclipse of the Sun and a total eclipse of the Moon each time the Moon orbits the Earth. The Earth, Moon, and Sun aren't always lined up precisely enough.

Lunar eclipses occur only when the Moon is full, which is when it is at its farthest distance from the Sun and is passing directly behind the Earth. The Moon is the right size and the right distance from Earth for the Earth's shadow to completely cover the Moon during a total lunar eclipse.

It so happens that the Moon is at just the right distance from Earth that it can fully block out the disk of the Sun during a total solar eclipse. Solar eclipses occur only during a new moon phase when the Moon is lined up precisely between the Sun and the Earth. Even though the disk of the Sun is covered by the Moon during a solar eclipse, intense sunlight shines around the edges of the Moon. If a total or partial solar eclipse is viewed directly, the intense sunlight can cause permanent damage to the eyes.

An interesting fact about the Moon is that the same side of the Moon always faces Earth. When we see a full Moon, the Moon is positioned so that the side facing Earth is illuminated, and during the new (or dark) Moon, this side is not illuminated. Because it takes the Moon about four weeks to orbit Earth, this means that at any particular place on the Moon, daylight lasts for two weeks and nighttime lasts for two weeks.

V. Just For Fun

Part A: Use your experimental setup to model how day and night occur on the Moon. Put a mark on the ping-pong ball to identify the side of the Moon that always faces Earth and then model the Moon's orbit.

Record your results.

Night and Day on the Moon

Part B: Think about how suns, planets, and moons might be arranged in different solar systems that we haven't yet visited. What do you think would happen with eclipses if you discovered a planet that, like Earth, had one moon, but the planet was orbiting two suns? What if a planet had two moons? What if the Sun were shining on the planet from above? What if Earth's Moon was a different shape, like a square or a dumbbell? What other different conditions can you think of? Model or draw some of your ideas and record your results in the following boxes.

Eclipses in Other Solar Systems

More Eclipses in Other Solar Systems

Experiment 4

Modeling the Moon

Introduction

Learn more about the Moon's features by building a model.

I. Think About It

❶ What do you think you can tell about the Moon's surface by looking at a photograph?

_____

_____

_____

_____

_____

❷ What similarities do you think there are between Earth and the Moon?

_____

_____

_____

_____

❸ What differences do you think there are between Earth and the Moon?

_____

_____

_____

_____

❹ Do you think it is possible for people to live on the Moon? Why or why not?

_____

_____

_____

_____

_____

❺ Do you think the first astronauts to walk on the Moon found anything surprising or things they did not already know? Why or why not?

_____

_____

_____

_____

❻ Do you think what scientists have learned about the Moon could help with planning for travel to other planets? Why or why not?

_____

_____

_____

_____

II. Experiment 4: Modeling the Moon   Date _____

Objective _____

_____

Hypothesis _____

_____

_____

Materials

    modeling clay in the following colors:
       gray
       white
       brown
       red
    butter knife or sculptor's knife
    ruler

EXPERIMENT

Model building is an important part of science. Models help scientists visualize how something might look in three dimensions.

❶ Observe the cutaway image and the photographs of the Moon in the *Student Textbook*.

❷ Using modeling clay, build a model Moon that resembles the cutaway image in the *Student Textbook*. Include the core, mantle, and crust and note the colors used for them. Observe any color variations on the Moon's surface. As much as possible, duplicate the cutaway image and photographs with your model.

❸ Use a ruler to measure the diameter of your completed model Moon.

Results

The real Moon is 3476.2 kilometers (2160 miles) in diameter. Compare the diameter of your model with the actual diameter of the Moon. Do the following steps to calculate how many times smaller your model is compared to the actual size of the Moon.

❶ Write the diameter of your model Moon in centimeters _____ or in inches

_____. The diameter of the actual Moon is 3476.2 kilometers (2160 miles).

❷ Convert the diameter of your model Moon to kilometers.

(If you are using inches, first multiply by 2.54 to get centimeters.

_____ inches X 2.54 = _____ centimeters.)

Multiply the number of centimeters by 0.00001 to get kilometers. This will be a very small number.

_____ centimeters X 0.00001 = _____ kilometers.

❸ Divide the actual diameter of the Moon by the diameter of your model Moon.

3476.2 kilometers (actual Moon) ÷ _____ kilometers (model Moon) =

_____.

This should be a very large number. It tells you how many times larger the real Moon is compared to your model Moon.

III. Conclusion

How easy or difficult was it to build a model of the Moon?

_____

_____

_____

_____

_____

_____

_____

Based on your calculation, how much larger is the actual Moon compared to your model Moon? What does this mean to you?

_____

_____

_____

_____

_____

_____

_____

_____

IV. Why?

What does the surface of the Moon look like and what is inside the Moon? Building a model is a great way to learn more about the Moon. To build a model, you need to look carefully at the features of the Moon that you can observe, and you also need to consider what scientists think about the structure of the interior part you can't see.

Suppose you wanted to add more detail to your model Moon than you can see in the photograph in your textbook. You might try using your eyes to look at the Moon on a clear night. You can see that it has different features that form dark and light areas, but you can't see the details. To see more detail you could look through binoculars. By looking through a telescope you'd be able to see much more. For an even closer look, you could view different photographs taken from satellites and other spacecraft and also photos and videos taken by astronauts who walked on the Moon. By studying photographs of the Moon, you might observe that it has lots of craters of all different sizes and does not appear to have vegetation or running water. The facts and details you gathered through observation could be added to your model.

Model building can help you compare what you've discovered about the Moon to what you know about Earth. In what ways are they similar and how are they different? You know that both the Earth and the Moon are celestial bodies and the Earth orbits the Sun. Because the Moon orbits Earth, the Moon travels along with Earth around the Sun. Both Earth and the Moon are spheres, are made of rock, have layers inside, and have rocky crusts that are not smooth. Because the Moon is smaller than Earth, it has less gravity. If you watch videos of the Apollo astronauts, you can see them bouncing as they walk on the Moon because the gravity is so much weaker than that of Earth.

The Moon has almost no atmosphere. Earth has a dense atmosphere that supports life. The Moon has ice but doesn't have surface water. Without an atmosphere and liquid water, the Moon can't support life as Earth does. Some astrobiologists think there's a possibility that microorganisms might exist on the Moon because bacteria and archaea that live in very extreme environments have been found on Earth and similar organisms may be able to survive on celestial bodies like the Moon.

V. Just For Fun

Imagine you are the designer of a permanent station on the Moon. What would the people need to live comfortably? What would they need for doing research? What would they do for fun? What else would they need? Write your ideas below, and on the next page draw a picture of your Moon station.

Ideas for a Moon Station

Station on the Moon

Experiment 5

Modeling the Planets

Introduction

Model the eight planets of our solar system to learn more about their similarities and differences.

I. Think About It

❶ If you were to make a list of some of Earth's features, what would they be?

_____

_____

_____

_____

_____

❷ Do you think anything can be learned by comparing the planets in our solar system? Why or why not?

_____

_____

_____

_____

_____

❸ Do you think all the Jovian planets are the same? Why or why not?

_____

_____

_____

_____

☆★☼○●☆★☼○●☆★☼○●☆★☼○●☆★☼○●☆★☼○●☆★☼○●☆★☼○●☆★☼○★

❹ Do you think a planet's position from the Sun determines any of the planet's features? Why or why not?

_____

_____

_____

_____

_____

❺ What methods and technologies do you think scientists use to find out more about the planets?

_____

_____

_____

_____

_____

❻ Do you think when people go to Mars they will find anything surprising? Why or why not?

_____

_____

_____

_____

_____

II. Experiment 5: Modeling the Planets   Date _____

Objective _____

_____

Hypothesis _____

_____

_____

Materials

Modeling clay in the following colors:
  gray
  white
  brown
  red
  blue
  green
  orange
butter knife or sculptor's knife
colored pencils

EXPERIMENT

❶ Look closely at the images of the eight planets in your *Student Textbook*. Observe their relative sizes (which planets are larger or smaller than the others) and their shape and colors.

❷ In the following spaces write notes about what you observe from the textbook images of each planet. Using colored pencils, make a quick sketch of each planet, noting any important features, such as rings or spots, and the colors of the features. These notes will be used as a guide for building your models. Online references can be used to find more images of the planets. You can go to a website such as www.nasa.gov/ or do a search for individual planet images.

Mercury

_____

_____

_____

_____

Venus

_____

_____

_____

Earth

_____

_____

_____

Mars

_____

_____

_____

_____

Jupiter

_____

_____

_____

_____

Saturn

_____

_____

_____

_____

Uranus

_____

_____

_____

_____

Neptune

_____

_____

_____

_____

❸ Using modeling clay, create a model of each planet. Refer to your notes and sketches while building the models, and make sure that you keep the relative sizes in proportion (Jupiter is larger than Earth, Mercury is smaller than Venus, and so on).

Results

Observe the model planets you have created. Are they the correct relative size? Do they match the images in the book? Are they all spherical in shape? Record your observations.

_____

_____

_____

_____

_____

_____

_____

_____

_____

_____

_____

_____

_____

_____

III. Conclusion

How easy or difficult was it to build the models of the planets?

_____

_____

_____

_____

_____

_____

_____

Discuss how well your models do or do not represent the real planets.

_____

_____

_____

_____

_____

_____

_____

IV. Why?

Model building can help you compare the planets in our solar system to each other, observing how they are similar and how they are different. To build models, you need to do research to find out details about each planet. By doing research you will find information you can use to build your models as accurately as possible. You will also find much additional information that you won't use in your models but that will help you understand much more about the planets.

Knowing that Mercury, Venus, Earth, and Mars are made of rock and have rocky crusts that are not smooth will help you build models of them. Knowing that Jupiter, Saturn, Uranus, and Neptune are made of gases can help with building those models. Finding out that the planets are spherical in shape, the size of each planet, which have rings, and what their surfaces look like will help you build more accurate models.

By doing research you can also discover many other facts about the planets that you won't include in your models. You can find out that the Earth and the other planets in our solar system each has its own orbit around the Sun and the distance of each from the Sun is known. You can find out what the atmosphere of each planet is made of and how thick it is, whether or not the planet has liquid water, and the temperature of the planet. You can learn how long the days and years are on a particular planet and how much its axis is tipped. These are just a few of the things you can discover.

We know that Earth has a thick atmosphere, liquid water, and is the right distance from the Sun to support life. By comparing what is known about Earth and the other planets, you can begin to develop ideas, or theories, about whether life might exist on another planet, whether we might someday travel to it, and whether we might eventually be able to set up a base on the planet where people could live and do research.

Doing research to find out what has already been discovered is a fun part of science because it can lead to finding information that is unexpected and even amazing! And doing this kind of research gives scientists the background information they need in order to be able to come up with new ideas for new discoveries.

V. Just For Fun

Imagine you are an astronomer using a powerful space telescope. You have just discovered a new planet, and the telescope brings detailed images to you. In the following box, name this new planet and list its features, such as what the planet is made of, what colors and patterns you can see, and whether it has water and life. Make a colored sketch of the planet and its features. If you'd like to see it in 3D, build a model of your new planet.

Planet _____

Experiment 6

Using a Star Map

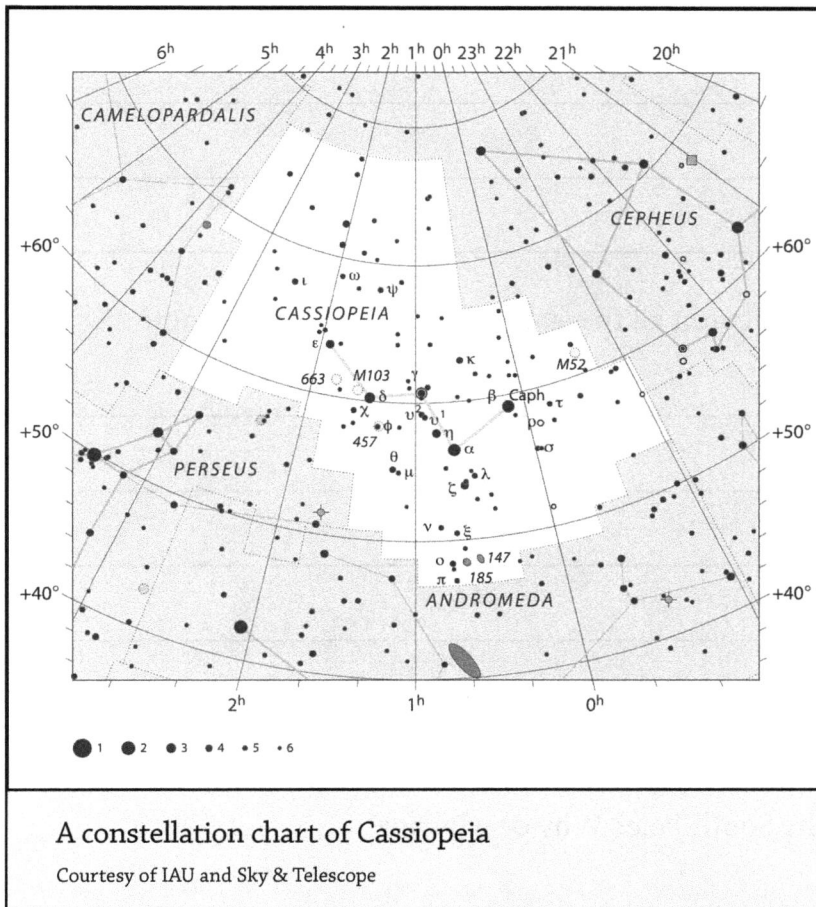

A constellation chart of Cassiopeia

Courtesy of IAU and Sky & Telescope

Introduction

Identify stars!

I. Think About It

❶ How many stars do you think are in the night sky? Why?

_____

_____

_____

_____

_____

❷ Do you think you could count all the stars you see? Why or why not?

_____

_____

_____

_____

_____

❸ Do you think you would see different stars in the night sky if you were at the North Pole than if you were at the South Pole? Why or why not?

_____

_____

_____

_____

_____

☆☆☼○★☆☼○★☆☆☼○★☆☼○★☆☼○★☆☼○★☆☆☼○★☆○☆

❹ If you had to map all the stars, how would you do it?

_____

_____

_____

_____

_____

❺ If you could travel back in time to the first century, do you think your star map would look different? Why or why not?

_____

_____

_____

_____

_____

❻ If you had an accurate map of the stars, in what ways might you use it?

_____

_____

_____

_____

II. Experiment 6: Using a Star Map   Date _____

Objective _____

_____

Materials

computer with internet access
printer and paper
flashlight

Optional

binoculars or telescope

EXPERIMENT

❶ Go to the Starmap site at http://www.star-map.fr/ and click on the Free Maps menu tab. This brings up a list of maps for the Northern Hemisphere. Under this list are three dots. Clicking on these will change the screen to show maps for the Equatorial Zone or the Southern Hemisphere.

Another free star map resource is http://www.skymaps.com.

[Note: Websites do change over time. If these sites are no longer available, do a browser search for "free star maps" to find a different resource.]

❷ On the Starmap site, select the hemisphere where you live, the map version you want to download, and the time you want to view the stars (20 pm is 8:00 pm and 22 pm is 10:00 pm). Click to download the map. Print the map.

❸ Study the star map you've downloaded. Check to make sure the map is for the correct hemisphere. Read the comments on the left side of the map. Many of the stars and constellations can be seen with an unaided eye, but some may require binoculars or a telescope.

❹ On an evening that is clear of clouds, go outside at the time you chose for the map you printed. Spend some time looking at the stars.

If you live in a city that has too much light at night for many stars to be seen, you may need to find a darker location away from city lights.

❺ Hold the map above your head so you are looking up at it and the sky. See if you can orient the map to match the stars you observe in the sky. Note which stars you are able to identify.

❻ In the space provided in the *Results* section, use the star map as a guide to make your own star map by recording the constellations you saw and the magnitude of the stars. Add planets or other objects and their location.

Results

Star Map   Date _____

III. Conclusion

A. Questions

❶ Based on your observations how easy was it to locate the constellations? Why?

_____

_____

_____

_____

❷ How many constellations were you able to record? Which ones?

_____

_____

_____

_____

❸ Were there any constellations you were not able to find? Why?

_____

_____

_____

_____

❹ How does artificial lighting affect your viewing of the stars?

_____

_____

_____

_____

☆☆☆☀○☆☆☆☀○☆☆☆☆☀○☆☆☀○☆☆☆☀○☆☆☆☀○☆☆☆☀○☆☆☆☀○☆

❺ If you were to view the stars for several hours, do you think you would be able to observe whether the constellations move? Why or why not?

_____

_____

_____

_____

B. Conclusions

Based on your observations, what conclusions can you draw?

_____

_____

_____

_____

_____

_____

_____

_____

_____

_____

_____

IV. Why?

By using a star map or star atlas you can become familiar with the stars. Like using a road map to find landmarks on Earth, you can learn landmarks in the night sky. If you were to study the night sky every night, you would become so familiar with it that you could navigate by the stars without using a star map.

There are many different kinds of star maps and star atlases available today. Some star maps show just the stars with the overlapping constellations. Some star maps have thousands of stars to observe and some only have a few hundred. There are also 3D star maps and planetary sphere maps that show star locations relative to your Earth position in a three-dimensional shape.

There are also deep space star maps that map the stars seen by the Hubble and Hipparcos telescopes. As astronomers continue to explore the cosmos, more stars and other celestial bodies are being added to star atlases.

V. Just For Fun

Repeat the experiment in a month. Download a new star map for that date and make your own star map based on the current location of the stars.

Compare your two star maps. What can you observe that is the same and what is different? The following pages provide space for your map and observations.

Star Map 2   Date _____

Star Map Comparisons

Star Map 1

Star Map 2

Experiment 7

Modeling Our Solar System

Introduction

Make a model of the planetary orbits of our solar system.

I. Think About It

❶ What do you think Earth would be like if it were in Mercury's orbit?

_____

_____

_____

_____

❷ Do you think life as we know it could exist on Jupiter? Why or why not?

_____

_____

_____

_____

❸ What do you think life on Earth would be like if it orbited two suns at the same time?

_____

_____

_____

_____

❹ What do you think it would be like if Earth's orbit was long and narrow instead of being almost round?

_____

_____

_____

_____

_____

❺ What do you think would happen if some planets orbited the Sun in a clockwise motion and others moved counterclockwise?

_____

_____

_____

_____

_____

❻ Why do you think planetary orbits are almost circular?

_____

_____

_____

_____

II. Experiment 7: Modeling Our Solar System   Date _____

Objective _____

_____

Hypothesis _____

_____

_____

Materials

8 objects of different sizes to represent the planets
ruler (in centimeters)
marker
large flat surface for drawing — 1 x 1 meter (3 x 3 feet), such as a
    large piece of cardboard or several sheets of construction paper
large open space at least 3 meters (10 feet) square
push pin
piece of string one meter (3 feet) long
tape

EXPERIMENT

❶ Find eight objects to represent the planets. Refer to the textbook illustration for the relative size of the planets and choose your objects to represent these sizes.

❷ Take the cardboard and mark the center with a marker. This represents the position of the Sun.

❸ Using the push pin, fix the string to the center mark of the cardboard.

❹ Measure 10 cm from the center and put a mark there. Wrap the loose end of the string around the marking pen so when the string is stretched out, the marking pen will be at the 10 cm mark. With the marker point touching the cardboard, draw a circle around the center mark. This is the orbital path for Earth.

☆★☼◐☆★☼◑○★☆☼◐○★☆☼◑○★☆☼◐○★☆☼◑○★☆☼◐○★☆☼◑○★☆☼◐○★☆☼◑○★

❺ Draw concentric circles for the first 5 planetary orbits (Mercury through Jupiter) using the distances listed below. You will need to adjust the length of the string for each orbit.

Planet	Distance from Center
Mercury	4 cm
Venus	7 cm
Earth	10 cm
Mars	15 cm
Jupiter	50 cm
Saturn	90 cm (3 ft)
Uranus	190 cm (6 ft)
Neptune	300 cm (10 ft)

❻ Place the objects you have chosen as your planetary models for the first 5 planets at their corresponding orbital distance from the center.

❼ For the last three orbits, measure the correct distance away from the center. Place the appropriate planetary model at the distance of its orbit.

Results

Observe your model of the solar system and compare it with the illustration in your *Student Textbook*. On the following page, note any similarities or differences between your model of the solar system and the illustration. What else can you observe about the solar system?

Similarities	Differences
_____	_____
_____	_____
_____	_____
_____	_____
_____	_____
_____	_____
_____	_____
_____	_____
_____	_____
_____	_____
_____	_____
_____	_____
_____	_____
_____	_____
_____	_____
_____	_____

☆☆☼◐○☆☆☼○☆☆☼○☆☆☼◐○☆☆☼○☆☆☼◐○☆☆☼○☆☆☼◐○☆☆☼○☆☆☆

III. Conclusion

How easy or difficult was it to create a model of the solar system? How did the different distances affect how you could build your model? What did you learn by building the model?

_____

_____

_____

_____

_____

_____

_____

_____

_____

_____

_____

_____

_____

_____

_____

IV. Why?

In this experiment you explored the orbital paths of the planets and how the planets are ordered in the solar system. An orbit is defined as the curved path that one celestial body follows as it travels around another celestial body. Although at one time it was thought that the Earth was the center of the universe and all the other celestial bodies orbited Earth, we now know that the Sun is the center of our solar system, making it a heliocentric system.

Basic physics tells us that bodies of mass have gravitational force, or gravity. The larger the body of mass, the more gravitational force it will have. The Sun is a very large body of mass and therefore has very strong gravitational force. Gravitational force keeps the planets in orbit around the Sun. The motion of the planets and the fact that the gravitational force of the Sun is constant are the things that keep the planetary orbits from collapsing towards the center of the solar system. The orbits of the planets are elliptical, but only slightly.

Mercury is in orbit closest to the Sun, and Neptune is farthest from the Sun. Measuring planetary distances is challenging because these distances are huge, and to show the distances in kilometers or miles results in very big numbers. To make it easier, astronomers use a unit of measure called the astronomical unit (AU) when talking about planetary distances. The distance from the Earth to the Sun is defined as 1 AU and the other planetary distances are some fraction or multiple of 1 AU. An AU is defined as the distance from Earth to the Sun because the distance of a planet from the Sun can be calculated using triangulation methods that require Earth's distance from the Sun as part of the calculation. Triangulation, or parallax, is still used as a method to arrive at distances and was used by ESA's Hipparcos satellite to accurately map the distances of over 100,000 stars. Radar and other methods are now also used to calculate distances.

The solar system can be divided into two different groups of planets according to their distance from the Sun. These groups are called the inner solar system and the outer solar system. There is a huge 4 AU gap between Mars (the outer planet of the inner solar system) and Jupiter (the inner planet of the outer solar system), and the Asteroid Belt is found in this gap.

V. Just For Fun

Expand the features of your solar system model.

Find additional items to add the Asteroid Belt to your model. Would there also be asteroids outside the Asteroid Belt? Would you see comets somewhere? Would you see moons or any artificial satellites orbiting any planets? Would you see any space probes or landers? What would they be looking for? What else might you add to your solar system model?

Expanded Solar System Model

Discovering Life
on Other Planets

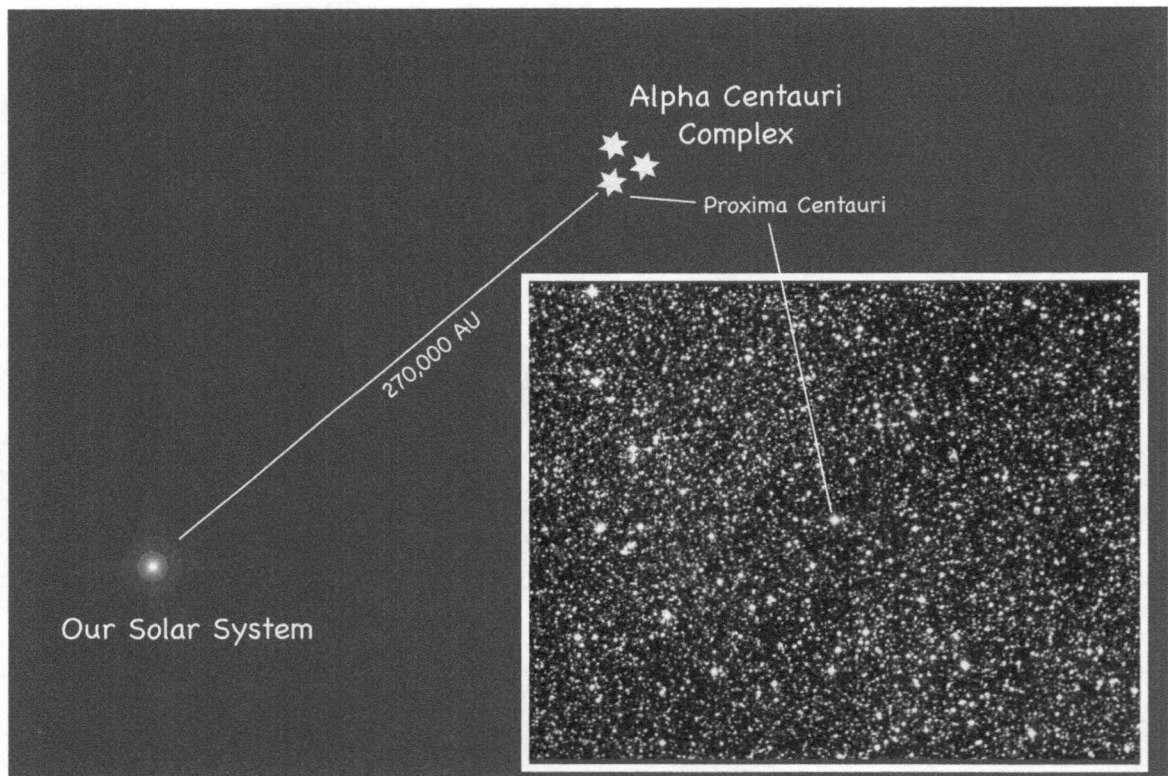

Alpha Centauri
Complex

Proxima Centauri

270,000 AU

Our Solar System

Introduction

Explore thought experiments.

I. Think About It

❶ What do you think is the likelihood that there is life on planets and/or moons outside our solar system? Why?

_____

_____

_____

_____

_____

❷ Do you think if life exists outside our solar system, it would be similar to life on Earth? Why or why not?

_____

_____

_____

_____

_____

❸ What factors do you think are necessary for life to exist on Earth? Why?

_____

_____

_____

_____

_____

❹ Do you think life outside our solar system would require the same conditions that we have on Earth? Why or why not?

_____

_____

_____

_____

_____

❺ How do you think you would detect "life" on another planet?

_____

_____

_____

_____

❻ How do you think living on a moon would be different from living on a planet? Why?

_____

_____

_____

_____

II. Experiment 8: Discovering Life on Other Planets

Date _____

Thought Experiment

Sometimes when it's not possible to do an actual experiment, it can be very useful to do what is called a *thought experiment*. A thought experiment is a mental exercise in which an experiment is imagined. The process of imagining how a hypothesis might be explored or how an experiment might actually work is very valuable to science. Albert Einstein wondered what it would be like to ride on a rainbow. He could not literally ride on a rainbow, but he could imagine it, and the ideas he generated during this thought experiment helped him discover the theory of relativity.

Materials

pencil
colored pencils
your imagination

EXPERIMENT

❶ Imagine that you are traveling outside our solar system and you come across a star three times the size of our Sun. You observe ten planets in the solar system around this sun. Some of the planets have moons. Assume that you can travel to all ten planets and explore all of their moons.

❷ Do a thought experiment and write in as much detail as possible what you would need to do to locate life on any of the ten planets or moons. Imagine this is really possible. Think about what you would need to take with you and how you would define "life." Also consider which planets or moons are more likely to have life and which you can ignore.

Discovering Life—A Thought Experiment

_____

_____

_____

_____

_____

_____

_____

_____

_____

_____

_____

_____

_____

_____

_____

_____

_____

_____

_____

_____

_____

_____

_____

_____

_____

_____

_____

_____

_____

_____

_____

_____

_____

_____

_____

_____

III. Why?

Because the universe is home to billions of stars, it makes sense to assume that some of those stars have planets. The idea of the existence of other planets has fascinated both scientists and science fiction writers for many years, but the existence of exoplanets has only recently been confirmed.

The idea that there might be life on other planets comes from the overwhelming number of possible planets that could orbit the billions of stars in the universe. Because we know the criteria for life on Earth, astronomers can begin to look at stars and the planets that orbit those stars to determine if there are any planets that meet the criteria for life as we know it.

Finding and studying exoplanets is extremely challenging. Most exoplanets lie close to their parent star. Direct imaging is difficult because the light from the star hides the planets. However, an exoplanet can be observed by direct imaging if the parent star is weakly luminous or if the exoplanet has a wide orbit.

Exoplanets can be indirectly observed by analyzing light from the parent stars. Recall that planets have mass and because of this have gravitational force. When a planet is orbiting a star, the star may "wobble" as a result of the planet's gravitational pull. The more massive the planet and the less massive the star, the more the star wobbles. Astronomers can use the wobbling of a star to estimate the mass of its exoplanets. Also, as a planet passes in front of the star it is orbiting, the light from the star dims slightly as the planet blocks some of the light. This slight dimming can be detected and used as another indirect method of finding exoplanets.

Although many exoplanets have been discovered, scientists are just beginning to find planets that might have the right conditions to be suitable to support life as we know it. It is thought that in order to support life, an exoplanet must be just the right distance from the parent star — neither too close nor too far away. This "Goldilocks distance" is called the Circumstellar Habitable Zone.

As technology advances, more and more discoveries will be made about exoplanets and the possibility that certain exoplanets could support life. The hope is that one day technology will advance enough that we can travel to distant planets to look for life and discover more about the universe we live in.

IV. Just For Fun

Review your thought experiment notes. In the space below make a diagram of the solar system you've explored, including the sun and the planets and any moons. Draw the orbits of the planets and name the solar system and the different celestial bodies in your diagram.

Diagram of the _____ **Solar System**

Choose one planet or moon where you imagine you have found life. Draw and/or write about the different life forms you found. How did you find them? What do they look like? What are they doing? What conditions do they need to live? (More space on next page.)

Life on (Planet or Moon) _____

Life Information (Continued)

Experiment 9

Astronomy Online

Courtesy of Gemini Observatory/AURA

Introduction

Access to astronomy data has expanded significantly because of the internet, allowing amateur astronomers to use online resources to explore sophisticated data. Here you will search the internet to find online tools for experiments in two following chapters.

I. Think About It

❶ What types of online programs have you used to explore astronomy?

_____

_____

_____

_____

❷ Have you ever used a star chart or online planetarium to look at planets in our solar system? If so, describe what you discovered.

_____

_____

_____

_____

❸ What other types of online resources have you used to explore any aspect of science? (For example, online chemistry charts, biology videos, physics demonstrations)

_____

_____

_____

_____

❹ How easy or difficult has it been to use online resources to learn about science?

_____

_____

_____

_____

_____

_____

❺ How reliable do you think the resources you've used are? How do you know the information you are viewing online is correct?

_____

_____

_____

_____

_____

❻ Do you use reviews or forums to discover what other resources people are using to learn about science online? Why or why not?

_____

_____

_____

_____

_____

II. Experiment 9: Astronomy Online

Date _____

Objective   _____

_____

Materials

computer
internet access

EXPERIMENT

❶ Read through Experiment 10, *The Center of the Milky Way,* and Experiment 12, *Searching for Nebulae.* You will need to collect specific information for these experiments. Make a short list of the information you will need to collect (e.g., nebulae, galaxies, galactic center, globular clusters, constellations, etc.).

_____

_____

_____

_____

_____

_____

_____

_____

_____

_____

_____

❷ Do an online search for astronomy software and websites. Try different keywords to find different types of resources, for example, "online astronomy," "astronomy programs," "astronomy software," etc. When you find a resource you might like to use, determine whether it will work on your computer. List the URLs for 3-5 resources along with a brief description of the type of information each provides.

_____

_____

_____

_____

_____

_____

_____

_____

_____

_____

_____

_____

_____

_____

_____

_____

❸ Review the resources listed in Step ❷. Which ones provide the information you are looking for? Do you need to download a program or can you use the program without downloading? Is it easy to use and understand? Does it allow you to search for the information you need, such as nebulae, galaxies, and the galactic center? Are the images easy to see? In the space below, make notes about the resources you've selected, including the types of information the program will provide for the remaining astronomy experiments.

_____

_____

_____

_____

_____

_____

_____

_____

_____

_____

_____

_____

_____

Results

Briefly describe the online resource(s) you have selected and why you think it will work for astronomy Experiments 10 and 12.

Astronomy Resource

_____

_____

_____

_____

_____

_____

_____

_____

_____

_____

_____

_____

_____

_____

_____

_____

III. Conclusion

What conclusions can you draw from your reserach?

_____

_____

_____

_____

_____

_____

_____

_____

_____

_____

_____

_____

_____

_____

_____

_____

IV. Why?

The internet can be a valuable resource for learning about astronomy. There are thousands of sites and programs available that contain star charts, images, and videos in addition to text. An amateur astronomer can find lots of fascinating information at levels everywhere from beginning to advanced.

In this experiment you researched astronomy software that can be used to find information for upcoming experiments. You discovered that there are several things that need to be considered when deciding upon an astronomy program to use for doing research. Can the program provide you with the data that will be needed? Is your computer compatible with the program's operating system requirements? Is the program understandable and easy to use? All of these factors play a role in determining the type of software you will need.

However, not all online resources are useful, and learning how to research science information is an important skill for using the internet effectively. For example, some sites require a user to download a program and install it on their computer. Some software requires a particular operating system, and if the computer you are using does not meet the program's requirements, the software cannot be used. Some software may be available only for purchase, which can be a limiting factor.

You may have discovered several online resources that seem like they would meet your needs, and you can use more than one resource to gather the data needed for the experiments that follow. In fact, it is good scientific practice to gather information from more than one resource whenever possible. Using more than one resource allows you to verify the information you are collecting. If you find inconsistencies between resources, you can try to determine which is the more reliable source, or you can check additional sources to see if the majority are in agreement about a fact. Because scientific study is complicated and scientists have differing theories and different experimental results, you may find that not everyone is in agreement. In looking for reliable resources, those that come from university sites are generally very reliable, as are sites that come from government agencies such as NASA, NOAA, and ESA (European Space Agency). Scientific journals can also be a good resource; however, care must be taken to ensure that these are fact-based publications with articles backed up by actual scientific study and data rather than just opinions.

The fun of doing internet research in astronomy is that it can lead you to the discovery of many amazing things in space that you didn't know existed. And with each new satellite, space telescope, space probe, lander, or rover that is launched, more and more discoveries are made that you can read about and see as images.

V. Just For Fun

NASA has many websites for its different missions and areas of research. Spend some time exploring various NASA sites—reading about NASA's missions and looking at the images and videos in the multimedia galleries. In the space provided make notes about the most interesting facts you find. Think about what information you would have to record in order to be able to quickly find a particular web page again and include this information in your notes. Also check out ESA websites. Do you find different information by going to space agencies from different countries?

Interesting Stuff from NASA and ESA

More Interesting Stuff from NASA and ESA

Experiment 10

The Center of the Milky Way

Artist's concept courtesy of ESO/NASA/JPL-Caltech/M. Kornmesser/R. Hurt

Introduction

Locate the center of the Milky Way galaxy using globular cluster data.

I. Think About It

❶ How easy or difficult do you think it would be to find the center of the Milky Way Galaxy?

_____

_____

_____

_____

❷ How would you describe a globular cluster?

_____

_____

_____

_____

❸ How do you think knowing about globular clusters could be helpful in finding the center of the Milky Way Galaxy?

_____

_____

_____

_____

❹ How many stars do you think a globular cluster contains? Why?

_____

_____

_____

_____

_____

_____

❺ What objects do you think are located in the galactic center of the Milky Way? How would you describe them?

_____

_____

_____

_____

_____

_____

❻ Do you think the galactic center can be observed from Earth? Why or why not?

_____

_____

_____

_____

_____

_____

II. Experiment 10: The Center of the Milky Way   Date _____

Objective _____

_____

Hypothesis _____

_____

Materials

computer
internet access

EXPERIMENT

❶ Set up the online resource you chose in the previous experiment.

❷ The *Appendix* at the back of this book gives data for globular clusters observed in our Milky Way Galaxy. The data table shows 158 globular clusters compiled as of June 30, 2010. From left to right the table lists the ID, name, and cross-reference for the cluster followed by the constellation where the cluster is located and various astronomical parameters associated with the cluster.

Look at the data table in the *Appendix*, and locate the three constellations that have the highest number of globular clusters. [Note: The number of globular clusters observed in a constellation is found in parentheses next to the constellation name. Constellations with fewer than two globular clusters are not listed.]

❸ In the chart below, record the three constellations that have the most globular clusters.

Constellation Name	# of Globular Clusters

Results

Open the resource you have chosen for finding the galactic center of the Milky Way Galaxy. Search for the three constellations listed in Step ❸ of the experiment. If your resource shows their locations, record this information. Since globular clusters are most numerous in the galactic center, these three constellations will lead you to the center of the Milky Way. Do a search for the location of the galactic center to check your results. In the space below, record your observations.

III. Conclusion

What conclusions can you draw from your research? Based on your observations, where is the galactic center of the Milky Way? How easy or difficult do you think it is to find the center of a galaxy?

_____

_____

_____

_____

_____

_____

_____

_____

_____

_____

_____

_____

_____

_____

IV. Why?

In this experiment you used and evaluated data on globular clusters to determine the center of the Milky Way galaxy. The appendix at the end of this workbook lists a number of different constellations together with the identification numbers, distance from the Sun or galactic center, apparent magnitude, and apparent dimension. The number of globular clusters in a constellation is found in parentheses next to the constellation name. Using this information you should have discovered that the globular clusters that contain the highest number of stars are Sagittarius with 34 globular clusters, Ophiuchus with 25 globular clusters, and Scorpio with 20 globular clusters. Because we know that the densest group of stars is at the center of the Milky Way Galaxy, we can use these three constellations with the largest number of globular clusters to find the galactic center.

Learning how to read and sort through scientific data is an important skill. Scientists often have to work with large amounts of data, sorting through numbers, names, and symbols. It takes time to learn how to study and evaluate scientific data. In this experiment the data was presented in a chart with the globular cluster count already identified. It might have taken longer or been more difficult if this information was not presented on a chart. You also may have noticed that some of the information listed on the chart was not used to determine the galactic center.

Depending on the software you selected, you should have been able to verify your results. You should have been directed to the identical location searching on the words "galactic center" as you found by typing in the three constellations, Sagittarius, Ophiuchus, and Scorpio. Being able to verify a result is an important step for any scientific research. If you were not able to verify the galactic center, you can redo the experiment or use a different software program. If the programs you are using are reliable, you should be able to verify your results.

The globular cluster chart includes information about right ascension, declination, and dimensions in arc minutes. With a little further research you can learn the meaning of these terms and how they are used in astronomy.

V. Just For Fun

Finding galaxies!

❶ Observe more galaxies with your astronomy software. Find the following galaxies, and in the spaces provided, draw what you observe.

Whirlpool Galaxy

NGC 1427A

M 101

M 82

Bode's Galaxy

M 87

Sombrero Galaxy

Sunflower Galaxy

Hoag's Object

Cartwheel Galaxy

NGC 3314

❷ Look for other galaxies. Find the ones that you think are the most interesting, beautiful, intricate, or weird and record what you see.

Galaxies!

More Galaxies!

Even More Galaxies!

Experiment 11

Classifying Galaxies

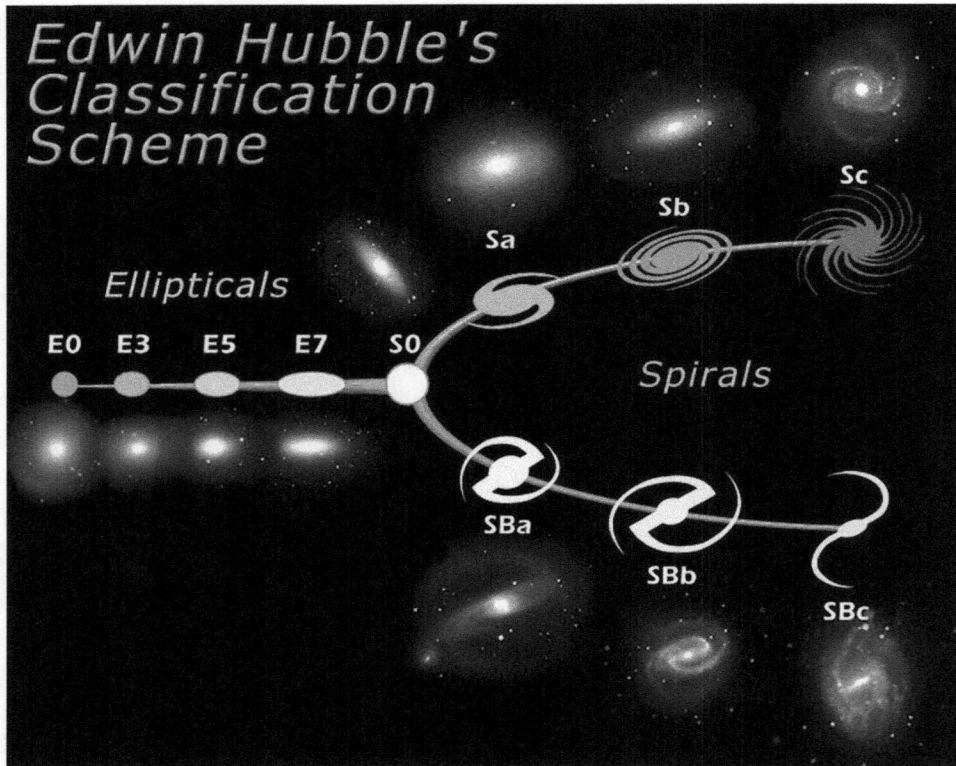

Edwin Hubble's Classification Scheme

Introduction

By performing this citizen science experiment, you'll be helping astronomers gather data for their science project while you learn more about classifying galaxies.

I. Think About It

❶ Why do you think it is useful for astronomers to put galaxies into groups?

_____

_____

_____

_____

❷ Do you think it's easy for astronomers to group galaxies according to how they look? Why or why not?

_____

_____

_____

_____

❸ If you were looking at images of galaxies, what characteristics do you think you might observe that would help you sort them into groups?

_____

_____

_____

_____

❹ Do you think astronomers have found all the different types of galaxies that exist? Why or why not?

_____

_____

_____

_____

_____

❺ What would you need to do to find out if a galaxy is a radio galaxy? Why?

_____

_____

_____

_____

_____

❻ If you discovered a new galaxy, what would you be most excited to find? Why?

_____

_____

_____

_____

_____

II. Experiment 11: Classifying Galaxies   Date _____

Objective   _____

_____

Hypothesis   _____

_____

Materials

computer or tablet
internet connection

EXPERIMENT

With all the powerful land- and space-based telescopes in use today, astronomers have an unimaginably huge number of images of celestial bodies that need to be looked at. In this experiment you'll classify galaxies to help the astronomers who are studying them.

❶   Go to the Galaxy Zoo website: www.galaxyzoo.org/

❷   Under the "Profile" tab, create a Zooniverse account by entering a username and a password. In your user account you will be able to view images of the galaxies you've classified. Clicking on one of these images will bring up more information about it.

❸   Explore the website by looking at the information in the dropdown menus under the different tabs on the menu bar at the top. The "Story" page has information about the project and also has links to the websites for telescopes used to collect the images.

❹   Go to the "Classify" page. Click on the box that says "Examples." Reviewing the examples is important to help you understand how to classify the galaxies.

❺   Before starting to classify galaxies, go to the menu tab "Discuss" and click on "Talk." Here you will find images that have been classified by others and comments about the features they identified in an image. This information may help you in making your own classifications, but you may or may not agree with their conclusions.

❻   Start classifying! Click on the box that best answers the question about the galaxy image being displayed. To get an opposite view of the image (dark on light), click on the image

or on the round blue button that says "Invert." This sometimes will reveal more features of the galaxy.

❼ Use the Results section or separate paper to keep notes as you go along.

Results

Galaxy Classification Notes

Type	Features

III. Conclusion

Based on your observations, how easy or difficult is it to classify galaxies? What did you discover about galaxies and images of them?

_____

_____

_____

_____

_____

_____

_____

_____

_____

_____

_____

_____

_____

_____

_____

_____

IV. Why?

In this experiment you probably noticed that galaxies can be hard to classify. Some galaxies are so far away that the galaxy looks like a fuzzy little blob, making it impossible to tell whether it has any features. It can also be difficult to tell the orientation of a galaxy. Is the image a side view, a top view, or is the galaxy tilted? Spiral arms, galactic bulges, and bars can be hard to distinguish. All of these factors can affect galaxy classification when using the Hubble Tuning Fork method.

Astronomers have come up with additional ways to classify galaxies. One method is to observe their color. Studying data that have been collected shows that spiral galaxies tend to be more blue because they have more active star forming areas and so have younger, hotter, blue stars. On the other hand, elliptical galaxies seem to have little ongoing star formation and so are made of older, cooler, red stars, giving the ellipticals their characteristic reddish color. Like other classification methods, this one is not perfect, and not all galaxies can be classified accurately by using it.

You may be wondering why blue stars are hotter and red stars are cooler when it seems like the opposite should be true. Recall from *Exploring the Building Blocks of Science Book 4* that in the optical range of the electromagnetic spectrum, blue has shorter wavelengths than red. In general, shorter wavelengths are more energetic than longer wavelengths. For example, imagine you and a friend are each holding one end of a long jump rope. If you jiggle one end of the rope slowly, the waves created will have long wavelengths. If you put a lot of energy into pumping the end of the rope up and down, you'll create short wavelength, high frequency waves. In a star, vibration of atoms and molecules creates the wavelength and frequency of the electromagnetic waves emitted. Heat provides energy, so hotter stars will have more rapidly vibrating atoms and molecules and will emit more energetic, shorter, higher frequency blue wavelengths. Cooler stars have less energy and less rapidly moving atoms and molecules and will emit less energetic, longer, lower frequency red waves.

But why classify galaxies at all? Astronomers have lots of questions with no easy answers. How did/do galaxies form? How long does it take? Is there star formation in all galaxies? Why? How? Do galaxies morph from one type to another? If so—how, why, when? How do they interact with each other? How do they move through space? Were galaxies that existed billions of years ago formed in the same way as galaxies that are only 2 million years old? How can we tell how old are they are, anyway? Where did everything in the universe come from? What is dark matter? What is dark energy? Is there really dark matter and dark energy? Scientists expect that classifying and studying galaxies will help answer these and many other questions. Lots of theories may come and go during the discovery process.

V. Just For Fun

❶ Go to Hubble Telescope website images: http://hubblesite.org/gallery/album/galaxy/

❷ Look at the images on the main "Galaxies" page (not the pages by class) and see if you can find one or more galaxies for each of the classes in the following chart.

❸ Click on the image to find out more about the galaxy. Fill in the chart with the name of the galaxy and interesting features. You can also print out your favorite images.

Galaxy Classifications

Spiral

Barred spiral

Galaxy Classifications

Elliptical

Lenticular

Irregular

Peculiar

Experiment 12

Searching for Nebulae

Courtesy of NASA, ESA and the Hubble SM4 ERO Team

Introduction

Use your online astronomy resource(s) to find nebulae.

I. Think About It

❶ How would you describe a nebula? Do you think there are different types of nebulae?

_____

_____

_____

_____

❷ Would we be able to study the universe if a black hole were in our solar system? Why or why not?

_____

_____

_____

_____

❸ Would we be able to study the universe if we lived on Jupiter? Why or why not?

_____

_____

_____

_____

❹ How might our study of the universe change if Earth were located in the middle of a nebula?

_____

_____

_____

_____

_____

_____

❺ How many nebulae do you think exists in the universe?

_____

_____

_____

_____

_____

_____

❻ How do you think a nebula might change over time?

_____

_____

_____

_____

_____

_____

II. Experiment 12: Searching for Nebulae   Date _____

Objective   _____

_____

Hypothesis   _____

_____

Materials

computer
internet access

EXPERIMENT

❶ Open the online resource you chose from Experiment 9, *Astronomy Online.*

❷ Search for the following nebulae:

Helix Nebula
Crab Nebula
Cone Nebula
Cat's Eye Nebula
Eagle Nebula
Orion Nebula

❸ In the *Results* section make a drawing of what you observe for each nebula and note the location of each by naming the constellation it is in and/or any nearby constellations. Find out what type of nebula you're looking at, what features it has, and any other information you can discover about it. Record your observations.

Results

Helix Nebula

Location and Information

_____

_____

_____

_____

_____

_____

Crab Nebula

Location and Information

_____

_____

_____

_____

_____

Cone Nebula

Location and Information

_____

_____

_____

_____

_____

_____

Cat's Eye Nebula

Location and Information

_____

_____

_____

_____

_____

_____

Eagle Nebula

Location and Information

_____

_____

_____

_____

_____

_____

Orion Nebula

Location and Information

_____

_____

_____

_____

_____

_____

III. Conclusion

What conclusions can you draw from your research?

_____

_____

_____

_____

_____

_____

_____

_____

_____

_____

_____

_____

_____

_____

_____

_____

_____

_____

IV. Why?

In this experiment you used your online resource to identify a variety of nebulae with their different shapes and features. As you learned in the *Student Textbook,* nebulae are composed of clouds of dust and gas, with the gas being mostly hydrogen. Some nebulae contain areas where stars and planetary systems form.

One type of nebula you observed is the diffuse emission nebula. The Cone Nebula and Orion Nebula are examples of diffuse emission nebulae. A diffuse nebula is a cloud of thin, widespread gas and dust particles. If a diffuse nebula is big and massive enough, it can have star forming regions. When young, massive, hot stars are forming in a diffuse nebula, their high energy radiation causes the gas around them (mostly hydrogen gas) to emit light, making the nebula shine. In this way a diffuse nebula becomes a diffuse emission nebula.

The Eagle Nebula is also a diffuse emission nebula and provides some of the most striking images of any nebula. The "Pillars of Creation" region of the nebula is a large area of active star formation and has long protruding columns and other strange and beautiful shapes.

The Helix Nebula, Crab Nebula, and Cat's Eye Nebula are planetary nebulae. Recall that a planetary nebula is formed when clouds of gas and dust are ejected from a red giant star as it becomes a white dwarf.

In images we see of nebulae, they seem to be composed of dense areas of gases, dust, and stars, but in reality the tiny particles nebulae are made of are extremely far apart. Nebulae are not dense at all, but because they are so far away, we observe them as gaseous clouds with many different and often oddly shaped features.

Astronomers spend hours and even years looking for nebulae and other objects in the night sky. Some nebulae can be observed with the naked eye, but the detail that we see today in images of nebulae is possible because of modern telescopes and technology, including the imaging of electromagnetic wavelengths that are not visible to the unaided eye. As more advanced telescopes and new technologies become available, we may be able to find many more nebulae and observe some that are even more distant or even less dense than those we can view now. We will also learn more about the composition of nebulae and how they form and change.

V. Just For Fun

Review the astronomy chapters you have studied. How many different types of objects described in the text can you find by using online resources? In the space provided, describe them in words and drawings.

Celestial Objects

More Celestial Objects

Appendix

Globular Clusters

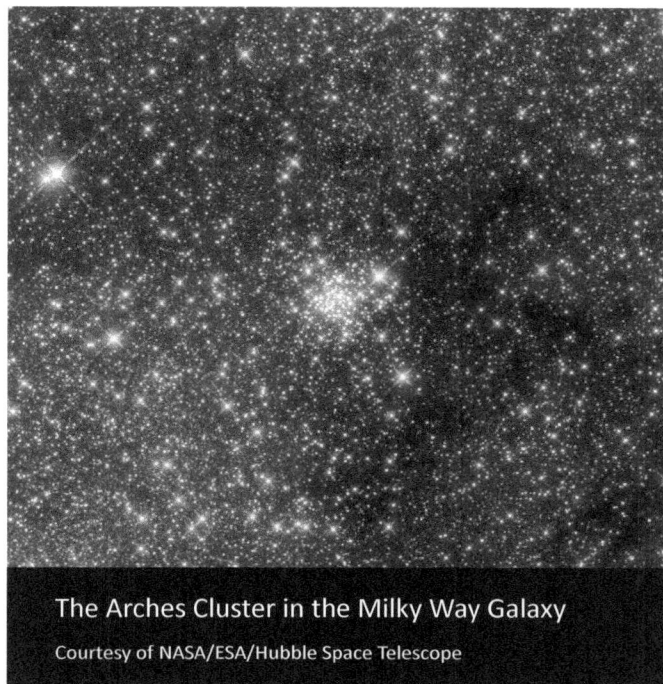

The Arches Cluster in the Milky Way Galaxy

Courtesy of NASA/ESA/Hubble Space Telescope

Appendix: Globular Clusters

On the following pages are several data tables with information about globular clusters observed in the Milky Way. Learning how to read data tables and sorting through the information to find the data needed for an experiment is an important part of scientific investigation.

Key

M, NGC/IC, ID/Name/Crossref

Messier, NGC, or IC number and other identification or name

Con

Constellation name (number of globular clusters in parentheses)

RA, Dec (2000)

Right Ascension and Declination for epoch 2000.0

R_Sun, R_gc

Distance from our Sun and the Galactic Center in thousands of light years (kly)

m_v

Apparent visual magnitude

dim

Apparent dimension in arc minutes

M	NGC/IC	ID/Name/Crossref	Con	RA (2000)	DEC	R_Sun	R_gc	m_v	dim
	104	47 Tuc Lac I.1	Tucana (2)	00:24:05.2	-72:04:51	14.7	24.1	3.95	50.0
	288	H 6.20	Scl	00:52:47.5	-26:35:24	28.7	39.1	8.09	13.0
	362	Dun 62	Tucana (2)	01:03:14.3	-70:50:54	27.7	30.3	6.40	14.0
		Whiting 1	Cet	02:02:56.8	-03:15:10				1.2
	1261	Dun 337	Horologium (2)	03:12:15.3	-55:13:01	53.5	59.4	8.29	6.6
		Pal 1	Cep	03:33:23.0	+79:34:50	35.6	55.4	13.18	2.8
		AM 1 E 1	Horologium (2)	03:55:02.7	-49:36:52	397.6	401.8	15.72	0.5
		Eri	Eri	04:24:44.5	-21:11:13	294.2	310.5	14.70	1.0
		Pal 2	Aur	04:46:05.9	+31:22:51	90.0	115.5	13.04	2.2
	1851	Dun 508	Col	05:14:06.3	-40:02:50	39.5	54.5	7.14	12.0
M 79	1904		Lep	05:24:10.6	-24:31:27	42.1	61.3	7.73	9.6
	2298	Dun 578	Pup	06:48:59.2	-36:00:19	34.9	51.2	9.29	5.0
	2419	H 1.218	Lyn	07:38:08.5	+38:52:55	274.6	298.4	10.39	4.6
		Koposov 2	Gem	07:58:17.0	+26:15:18	130			
		Pyxisis	Pyx	09:07:57.8	-37:13:17	129.4	135.9	12.90	4.0
	2808	Dun 265	Car	09:12:02.6	-64:51:47	31.2	36.2	6.20	14.0
		E 3	Cha	09:20:59.3	-77:16:57	14.0	24.8	11.35	10:
		Pal 3	Sex	10:05:31.4	+00:04:17	302.3	312.8	14.26	1.6
		Segue 1	Leo	10:07:04	+12:47:30	75.0	14.7	4.5	
	3201	Dun 445	Vel	10:17:36.8	-46:24:40	16.3	29.0	6.75	20.0
		Pal 4	UMa	11:29:16.8	+28:58:25	356.2	364.6	14.20	1.3
		Koposov 1	Virgo (2)	11:59:18.5	+12:15:36	160			
	4147	H 1.19	Coma Berenices (3)	12:10:06.2	+18:32:31	62.9	69.5	10.32	4.4
	4372		Mus	12:25:45.4	-72:39:33	18.9	23.2	7.24	5.0
		Rup 106	Cen	12:38:40.2	-51:09:01	69.1	60.3	10.90	2.0
M 68	4590		Hya	12:39:28.0	-26:44:34	33.3	32.9	7.84	11.0
	4833	Lac I.4 Dun 164	Mus	12:59:35.0	-70:52:29	21.2	22.8	6.91	14.0
M 53	5024		Coma Berenices (3)	13:12:55.3	+18:10:09	58.0	59.6	7.61	13.0
	5053	H 6.7	Coma Berenices (3)	13:16:27.0	+17:41:53	53.5	55.1	9.47	10.0
	5139	Omega Cen Lac I.5	Cen	13:26:45.9	-47:28:37	17.3	20.9	3.68	55.0
M 3	5272		CVn	13:42:11.2	+28:22:32	33.9	39.8	6.19	18.0
	5286	Dun 388	Cen	13:46:26.5	-51:22:24	35.9	27.4	7.34	11.0
		AM 4	Hya	13:56:21.2	-27:10:04	97.5	83.2	15.90	3.0
	5466	H 6.9	Boo	14:05:27.3	+28:32:04	51.8	52.8	9.04	9.0
	5634	H 1.70	Virgo (2)	14:29:37.3	-05:58:35	82.2	69.1	9.47	5.5
	5694	H 2.196	Hya	14:39:36.5	-26:32:18	113.2	94.9	10.17	4.3

		I4499	Aps	15:00:18.5	-82:12:49	61.6	51.2	9.76	8.0
	5824		Lup	15:03:58.5	-33:04:04	104.4	84.1	9.09	7.4
		Pal 5	SerCp	15:16:05.3	-00:06:41	75.7	60.7	11.75	8.0
	5897	H 6.8 H 6.19	Lib	15:17:24.5	-21:00:37	40.4	23.8	8.53	11.0
M 5	5904		SerCp	15:18:33.8	+02:04:58	24.5	20.2	5.65	23.0
	5927	Dun 389	Lup	15:28:00.5	-50:40:22	24.8	14.7	8.01	6.0
	5946		Norma (3)	15:35:28.5	-50:39:34	34.6	18.9	9.61	3.0
		BH 176	Norma (3)	15:39:07.3	-50:03:02	50.9	31.6	14.00	3.0
	5986	Dun 552	Lup	15:46:03.5	-37:47:10	33.9	15.7	7.52	9.6
		Lynga 7	Norma (3)	16:11:03.0	-55:18:52	23.5	2.5		
		Pal 14 AvdB	Her	16:11:04.9	+14:57:29	241.0	225.0	14.74	2.5
M 80	6093		Scorpio (20)	16:17:02.5	-22:58:30	32.6	12.4	7.33	10.0
M 4	6121	Lac I.9	Scorpio (20)	16:23:35.5	-26:31:31	7.2	19.2	5.63	36.0
	6101	Dun 68	Aps	16:25:48.6	-72:12:06	49.9	36.2	9.16	5.0
	6144	H 6.10	Scorpio (20)	16:27:14.1	-26:01:29	27.7	8.5	9.01	7.4
	6139	Dun 536	Scorpio (20)	16:27:40.4	-38:50:56	32.9	11.7	8.99	8.2
		Terzan 3	Scorpio (20)	16:28:40.1	-35:21:13	24.5	7.8	12.00	3.0
M 107	6171	H 6.40	Ophiuchus (25)	16:32:31.9	-13:03:13	20.9	10.8	7.93	13.0
		1636-283 ESO452-SC11	Scorpio (20)	16:39:25.5	-28:23:52	25.4	6.5	12.00	1.2
M 13	6205	H 4.50	Her	16:41:41.5	+36:27:37	25.1	28.4	5.78	20.0
	6229		Her	16:46:58.9	+47:31:40	99.1	96.8	9.39	4.5
M 12	6218		Ophiuchus (25)	16:47:14.5	-01:56:52	16.0	14.7	6.70	16.0
		FSR 1735 2MASS-GC03	Arae (5)	16:52:10.6	-47:03:29	29.7	10.4	0.8	
	6235	H 2.584	Ophiuchus (25)	16:53:25.4	-22:10:38	37.2	13.4	9.97	5.0
M 10	6254		Ophiuchus (25)	16:57:08.9	-04:05:58	14.4	15.0	6.60	20.0
	6256		Scorpio (20)	16:59:32.6	-37:07:17	27.4	5.9	11.29	4.1
		Pal 15	Ophiuchus (25)	17:00:02.4	-00:32:31	145.5	123.6	14.00	3.0
M 62	6266	Dun 627	Ophiuchus (25)	17:01:12.6	-30:06:44	22.5	5.5	6.45	15.0
M 19	6273		Ophiuchus (25)	17:02:37.7	-26:16:05	28.0	5.2	6.77	17.0
	6284	H 6.11	Ophiuchus (25)	17:04:28.8	-24:45:53	49.9	24.8	8.83	6.2
	6287	H 2.195	Ophiuchus (25)	17:05:09.4	-22:42:29	30.3	6.8	9.35	4.8
	6293	H 6.12	Ophiuchus (25)	17:10:10.4	-26:34:54	28.7	4.6	8.22	8.2
	6304	H 1.147	Ophiuchus (25)	17:14:32.5	-29:27:44	19.6	7.2	8.22	8.0
	6316	H 1.45	Ophiuchus (25)	17:16:37.4	-28:08:24	35.9	10.4	8.43	5.4
M 92	6341		Her	17:17:07.3	+43:08:11	26.7	31.3	6.44	14.0
	6325		Ophiuchus (25)	17:17:59.2	-23:45:57	26.1	3.6	10.33	4.1
M 9	6333		Ophiuchus (25)	17:19:11.8	-18:30:59	25.8	5.5	7.72	12.0
	6342	H 1.149	Ophiuchus (25)	17:21:10.2	-19:35:14	28.0	5.5	9.66	4.4
	6356	H 1.48	Ophiuchus (25)	17:23:35.0	-17:48:47	49.6	24.8	8.25	10.0

6355	H 1.46	Ophiuchus (25)	17:23:58.6	-26:21:13	31.0	5.9	9.14	4.2
6352	Dun 417	Arae (5)	17:25:29.2	-48:25:22	18.6	10.8	7.96	9.0
	I1257	Ophiuchus (25)	17:27:08.5	-07:05:35	81.5	58.4	13.10	5.0
	Terzan 2 HP 3	Scorpio (20)	17:27:33.4	-30:48:08	28.4	2.9	14.29	0.6
6366		Ophiuchus (25)	17:27:44.3	-05:04:36	11.7	16.3	9.20	13.0
	Terzan 4 HP 4	Scorpio (20)	17:30:38.9	-31:35:44	29.7	4.2	16.00	0.7
	HP 1 BH 229	Ophiuchus (25)	17:31:05.2	-29:58:54	46.0	19.9	11.59	1.2
6362	Dun 225	Arae (5)	17:31:54.8	-67:02:53	24.8	16.6	7.73	15.0
	Liller 1	Scorpio (20)	17:33:24.5	-33:23:20	34.2	8.5	16.77	12.6
6380	Ton 1	Scorpio (20)	17:34:28.0	-39:04:09	34.9	10.4	11.31	3.6
	FSR 1767	Scorpio (20)	17:35:43	-36:21:28	4.9	18.6		
	Terzan 1 HP 2	Scorpio (20)	17:35:47.8	-30:28:11	18.3	8.2	15.90	2.4
	Ton 2 Pismis 26	Scorpio (20)	17:36:10.5	-38:33:12	26.4	4.6	12.24	2.2
6388	Dun 457	Scorpio (20)	17:36:17.0	-44:44:06	32.6	10.4	6.72	10.4
M 14 6402	H 1.44	Ophiuchus (25)	17:37:36.1	-03:14:45	30.3	13.4	7.59	11.0
6401	H 1.44	Ophiuchus (25)	17:38:36.9	-23:54:32	34.2	8.8	9.45	4.8
6397	Lac III.11 Dun 366	Arae (5)	17:40:41.3	-53:40:25	7.5	19.6	5.73	31.0
	Pal 6	Ophiuchus (25)	17:43:42.2	-26:13:21	19.2	7.2	11.55	1.2
6426	H 2.587	Ophiuchus (25)	17:44:54.7	+03:10:13	67.5	47.6	11.01	4.2
	Djorg 1	Scorpio (20)	17:47:28.3	-33:03:56	39.1	13.4	13.60	
	Terzan 5 Terzan 11	Sagittarius (34)	17:48:04.9	-24:48:45	33.6	7.8	13.85	2.4
6440	H 1.150	Sagittarius (34)	17:48:52.6	-20:21:34	27.4	4.2	9.20	4.4
6441	Dun 557	Scorpio (20)	17:50:12.9	-37:03:04	38.1	12.7	7.15	9.6
	Terzan 6 HP 5	Scorpio (20)	17:50:46.4	-31:16:31	31.0	5.2	13.85	1.4
6453		Scorpio (20)	17:50:51.8	-34:35:55	31.3	5.9 10.0	7.6	
	UKS 1 UKS 1751-241	Sagittarius (34)	17:54:27.2	-24:08:43	27.1	2.6	17.29	2.0
6496	Dun 460	Scorpio (20)	17:59:02.0	-44:15:54	37.5	14.0	8.54	5.6
	Terzan 9	Sagittarius (34)	18:01:38.8	-26:50:23	21.2	5.2	16.00	0.2
	Djorg 2 E456-SC38	Sagittarius (34)	18:01:49.1	-27:49:33	21.9	4.6	9.90	9.9
6517	H 2.199	Ophiuchus (25)	18:01:50.6	-08:57:32	35.2	14.0	10.23	4.0
	Terzan 10	Sagittarius (34)	18:02:57.4	-26:04:00	18.6	7.8	14.90	1.5
6522	H 1.49	Sagittarius (34)	18:03:34.1	-30:02:02	25.4	2.0	8.27	9.4
6535		SerCd	18:03:50.7	-00:17:49	22.2	12.7	10.47	3.4
6528	H 2.200	Sagittarius (34)	18:04:49.6	-30:03:21	25.8	2.0	9.60	5.0
6539		SerCd	18:04:49.8	-07:35:09	27.4	10.1	9.33	7.9
6540	H 2.198 Djorg 3	Sagittarius (34)	18:06:08.6	-27:45:55	12.1	14.4	9.30	1.5
6544	H 2.197	Sagittarius (34)	18:07:20.6	-24:59:51	8.8	17.3	7.77	9.2
6541	Dun 473	CrA	18:08:02.2	-43:42:20	22.8	7.2	6.30	15.0
	2MASS-GC01	Sagittarius (34)	18:08:21.8	-19:49:47	11.7	14.7	3.3	

M	NGC	Name	Constellation	RA	Dec				
		ESO 280-SC06	Arae (5)	18:09:06	-46:25:24	70.7	46.6	1.5	
	6553	H 4.12	Sagittarius (34)	18:09:15.6	-25:54:28	19.6	7.2	8.06	9.2
		2MASS-GC02	Sagittarius (34)	18:09:36.5	-20:46:44	13.0	13.4	1.9	4.2
	6558		Sagittarius (34)	18:10:18.4	-31:45:49	24.1	3.3	9.26	8.0
		I1276 Pal 7	SerCd	18:10:44.2	-07:12:27	17.6	12.1	10.34	1.0
		Terzan 12	Sagittarius (34)	18:12:15.8	-22:44:31	15.7	11.1	15.63	6.4
	6569	H 2.201 Dun 619	Sagittarius (34)	18:13:38.9	-31:49:35	34.9	9.5	8.55	1.3
		AL 3	Sagittarius (34)	18:14:05.7	-28:38:08				
	6584	Dun 376	Tel	18:18:37.7	-52:12:54	43.7	22.8	8.27	6.6
	6624	H 1.50	Sagittarius (34)	18:23:40.5	-30:21:40	25.8	3.9	7.87	8.8
M 28	6626	Lac I.11	Sagittarius (34)	18:24:32.9	-24:52:12	18.3	8.8	6.79	11.2
	6638	H 1.51	Sagittarius (34)	18:30:56.2	-25:29:47	31.2	7.5	9.02	7.3 M,
	6637	Lac I.12 Dun 613	Sagittarius (34)	18:31:23.2	-32:20:53	29.7	6.2	7.64	9.8
	6642	H 2.205	Sagittarius (34)	18:31:54.3	-23:28:35	27.4	5.5	9.13	5.8
	6652		Sagittarius (34)	18:35:45.7	-32:59:25	32.9	9.1	8.62	6.0
M 22	6656		Sagittarius (34)	18:36:24.2	-23:54:12	10.4	16.0	5.10	32.0
		Pal 8	Sagittarius (34)	18:41:29.9	-19:49:33	42.1	18.3	11.02	5.2
M 70	6681	Dun 614	Sagittarius (34)	18:43:12.7	-32:17:31	29.4	6.8	7.87	8.0
		GLIMPSE-C01	Aquila (4)	18:48:49.7	-01:29:50			10-17	
	6712	H 1.47	Sct	18:53:04.3	-08:42:22	22.5	11.4	8.10	9.8
M 54	6715	Dun 624	Sagittarius (34)	18:55:03.3	-30:28:42	87.3	62.6	7.60	12.0
	6717	H 3.143 Pal 9	Sagittarius (34)	18:55:06.2	-22:42:03	23.1	7.8	9.28	5.4
	6723	Dun 573	Sagittarius (34)	18:59:33.2	-36:37:54	28.4	8.4	7.01	13.0
	6749	Berkeley 42	Aquila (4)	19:05:15.3	+01:54:03	25.8	16.3	12.44	4.0
	6752	Dun 295	Pav	19:10:51.8	-59:58:55	13.0	17.0	5.40	29.0
	6760		Aquila (4)	19:11:12.1	+01:01:50	24.1	15.7	8.88	9.6
M 56	6779		Lyr	19:16:35.5	+30:11:05	32.9	31.6	8.27	8.8
		Terzan 7	Sagittarius (34)	19:17:43.7	-34:39:27	75.7	52.2	12.00	1.2
		Pal 10	Sge	19:18:02.1	+18:34:18	19.2	20.9	13.22	4.0
		Arp 2	Sagittarius (34)	19:28:44.1	-30:21:14	93.3	69.8	12.30	2.3
M 55	6809	Lac I.14 Dun 620	Sagittarius (34)	19:39:59.4	-30:57:44	17.3	12.7	6.32	19.0
		Terzan 8	Sagittarius (34)	19:41:45.0	-34:00:01	84.8	62.3	12.40	3.5
		Pal 11	Aquila (4)	19:45:14.4	-08:00:26	42.4	25.8	9.80	10.0
M 71	6838		Sge	19:53:46.1	+18:46:42	13.0	21.9	8.19	7.2
M 75	6864		Sagittarius (34)	20:06:04.8	-21:55:17	67.5	47.6	8.52	6.8
	6934	H 1.103	Delphinus (2)	20:34:11.6	+07:24:15	51.2	41.7	8.83	7.1
M 72	6981		Aqr	20:53:27.9	-12:32:13	55.4	42.1	9.27	6.6
	7006	H 1.52	Delphinus (2)	21:01:29.5	+16:11:15	135.4	126.5	10.56	3.6
M 15	7078		Peg	21:29:58.3	+12:10:01	33.6	33.9	6.20	18.0

M	NGC	Name	Constellation	RA	Dec				
M 2	7089		Aqr	21:33:29.3	-00:49:23	37.5	33.9	6.47	16.0
M 30	7099		Cap	21:40:22.0	-23:10:45	26.1	23.2	7.19	12.0
		Pal 12 Cap Dwarf	Cap	21:46:38.8	-21:15:03	62.3	51.9	11.99	2.9
		Pal 13	Peg	23:06:44.4	+12:46:19	84.1	87.0	13.80	0.7
	7492	H 3.558	Aqr	23:08:26.7	-15:36:41	84.1	81.2	11.29	4.2

References

V. Berokurov, D. B. Zucker, N. W. Evans, J. T. Kleyna, S. Koposov, S. T. Hodgkin, M. J. Irwin, G. Gilmore, M. I. Wilkinson, M. Fellhauer, D. M. Bramich, P. C. Hewett, S. Vidrir, J. T. A. de Jong, J. A. Smith, H.-W. Rix, E. F. Bell, R. F. G. Wyse, H. J. Newberg, P. A. Mayeur, B. Yanny, C. M. Rockosi, O. Y. Gnedin, D. P. Schneider, T. C. Beers, J. C. Barentine, H. Brewington, J. Brinkmann, M. Harvanek, S. J. Kleinman, J. Krzesinski, D. Long, A. Nitta, S. A. Sneddon, 2007. Cats and Dogs, Hair and a Hero: A Quintet of New Milky Way Companions. Astrophysical Journal, Vol. 654, Issue 2, pp. 897-906 (January 2007) [ADS: 2007ApJ...654..897B] - [Preprint: astro-ph/0608448] Discovery announce of Segue 1.

C. Bonatto, E. Bica, S. Ortolani and B. Barbuy, 2007. FSR 1767 - a new globular cluster in the Galaxy. To be published in: Monthly Notices of the Royal Astronomical Society. [Preprint: arXiv:0708.0501[astro-ph]] Discovery announce of FSR-1767.

D. Froebrich, H. Meusinger and A. Scholz, 2007. SR 1735 - A new globular cluster candidate in the inner Galaxy. To appear in: Monthly Notices of the Royal Astronomical Society (2007). [Preprint: astro-ph/0703318] Discovery announce of FSR-1735.

W. E. Harris, 1996-1999. Catalog of Parameters for Milky Way Globular Clusters. AJ, 112, 1487. Revision of June 22, 1999. Available online; also see references and the potentially more current original site.

R. J. Hurt, et al., 2000. Serendipitous 2MASS Discoveries near the Galactic Plane: A Spiral Galaxy and Two Globular Clusters. The Astronomical Journal, Volume 120, Issue 4, pp. 1876-1883 (10/2000). [ADS: 2000AJ....120.1876H] - [Preprint] Discovery announce of two new globulars, 2MASS-GC01 and 2MASS-GC02.

Henry Kobulnicky, B. L. Babler, T. M. Bania, R. A. Benjamin, B. A. Buckalew, R. Canterna, E. Churchwell, D. Clemens, M. Cohen, J. M. Darnel, J. M. Dickey, R. Indebetouw, J. M. Jackson, A. Kutyrev, A. P. Marston, J. S. Mathis, M. R. Meade, E. P. Mercer, A. J. Monson, J. P. Norris, M. J. Pierce, R. Shah, J. R. Stauffer, S. R. Stolovy, B. Uzpen, C. Watson, B. A. Whitney, M. J. Wolff, and M. G. Wolfire, 2004. Newfound Star Cluster may be final Milky Way 'Fossil.' Spitzer Science Center News Release 2004-16. Discovery announce of the new globular GLIMPSE-C01.

S. Koposov, J. T. A. de Jong, H.-W. Rix, D. B. Zucker, N. W. Evans, G. Gilmore, M. J. Irwin, E. F. Bell, 2007. The discovery of two extremely low luminosity Milky Way globular clusters. Submitted to Astrophysical Journal. [Preprint: arXiv:0706.0019[astro-ph]] Discovery paper of Koposov 1 and Koposov 2.

S. Ortolani, E. Bica and B. Barbuy (2000). ESO 280-SC06: a new globular cluster in the Galaxy. Astronomy and Astrophysics, Vol. 361, pp. L57-L59 (September 2000). [ADS: 2000A&A...361L..57O] Discovery announce of the new globular, ESO 280-SC06.

S. Ortolani, E. Bica and B. Barbuy, 2006. AL 3 (BH 261): a new globular cluster in the Galaxy. Astrophysical Journal, Vol. 646, Issue 2, pp. L115-L118 (August 2006) [ADS: 2006ApJ...646L.115O] - [Preprint: astro-ph/0606718]. Discovery announce of the globular nature of AL-3.

More REAL SCIENCE-4-KIDS Books
by Rebecca W. Keller, PhD

Building Blocks Series
yearlong study program — each Student Textbook has accompanying Laboratory Notebook, Teacher's Manual, Lesson Plan, Study Notebook, Quizzes, and Graphics Package

Exploring the Building Blocks of Science Book K (Activity Book)
Exploring the Building Blocks of Science Book 1
Exploring the Building Blocks of Science Book 2
Exploring the Building Blocks of Science Book 3
Exploring the Building Blocks of Science Book 4
Exploring the Building Blocks of Science Book 5
Exploring the Building Blocks of Science Book 6
Exploring the Building Blocks of Science Book 7
Exploring the Building Blocks of Science Book 8

Focus Series
unit study program — each title has a Student Textbook with accompanying Laboratory Notebook, Teacher's Manual, Lesson Plan, Study Notebook, Quizzes, and Graphics Package

Focus On Elementary Chemistry
Focus On Elementary Biology
Focus On Elementary Physics
Focus On Elementary Geology
Focus On Elementary Astronomy

Focus On Middle School Chemistry
Focus On Middle School Biology
Focus On Middle School Physics
Focus On Middle School Geology
Focus On Middle School Astronomy

Focus On High School Chemistry

Super Simple Science Experiments

21 Super Simple Chemistry Experiments
21 Super Simple Biology Experiments
21 Super Simple Physics Experiments
21 Super Simple Geology Experiments
21 Super Simple Astronomy Experiments
101 Super Simple Science Experiments

Note: A few titles may still be in production.

Gravitas Publications Inc.
www.gravitaspublications.com
www.realscience4kids.com

GRAVITAS
PUBLICATIONS

www.ingramcontent.com/pod-product-compliance
Lightning Source LLC
Chambersburg PA
CBHW080559220326
41599CB00032B/6542